# Human Factors Applications in Teleoperator Design and Operation

WILEY SERIES IN HUMAN FACTORS

Edited by David Meister

**Human Factors in Quality Assurance**
Douglas H. Harris and Frederick B. Chaney

**Man-Machine Simulation Models**
Arthur I. Siegel and J. Jay Wolf

**Human Factors Applications in Teleoperator Design and Operation**
Edwin G. Johnsen and William R. Corliss

**Human Factors: Theory and Practice**
David Meister

# Human Factors Applications in Teleoperator Design and Operation

Edwin G. Johnsen and William R. Corliss

Wiley-Interscience, A Division of John Wiley & Sons, Inc.
New York · London · Sydney · Toronto

Copyright © 1971, by John Wiley & Sons, Inc.

All rights reserved. Published simultaneously in Canada.

No part of this book may be reproduced by any means, nor transmitted, nor translated into a machine language without the written permission of the publisher.

Library of Congress Catalog Card Number: 79-137107

ISBN 0 471 44292 5

Printed in the United States of America

10 9 8 7 6 5 4 3 2 1

# FOREWORD

In pursuing their objectives, both the Atomic Energy Commission and the National Aeronautics and Space Administration must continue to operate in the forefront of science and technology. New challenges must be faced and new tools developed to permit us to further extend our capabilities to meet future human needs. In pursuing advanced programs dealing with space and the atom man has recognized his current limitations and is applying new knowledge to overcome them. One of the most stimulating and rewarding areas of development has been in the field of teleoperators—man-machine systems that augment man by projecting and magnifying his manipulatory capability into inaccessible environments.

Our agencies are making increasing use of these man-machine combinations. We are proud of the advances in teleoperator design and operation that have already been achieved, but recognize that in this new field many accomplishments are yet to be realized. Everyone is now familiar with various types of teleoperators, ranging from simple kitchen tongs to artificial limbs to an unmanned surveyor spacecraft digging a trench in the lunar soil. More sophisticated teleoperators are under development which the society of the future will utilize even more heavily to extend and transform the intellectual and physical capacities of man.

Edwin G. Johnsen and William R. Corliss, experts in teleoperator theory, design, and application, have captured the movement and excitement of the field of human augmentation in this book. While aimed primarily at those generally unfamiliar with teleoperators, the book also inspires the reader to let his imagination carry him into the future when man-machine combinations will become even more commonplace. We commend the authors on this important work.

        Glenn T. Seaborg, Chairman
        *U. S. Atomic Energy Commission*
        T. O. Paine, Administrator
        *National Aeronautics and Space Administration*

# SERIES PREFACE

Technology is effective to the extent that men can operate and maintain the machines they design. Equipment design which consciously takes advantage of human capabilities and constrains itself within human limitations amplifies and increases system output. If it does not, system performance is reduced and the purpose for which the equipment was designed is endangered. This consideration is even more significant today than in the past because the highly complex systems that we develop are pushing human functions more and more to their limits of efficient performance.

How can one ensure that machines and machine operations are actually designed for human use? Behavioral data, principles, and recommendations—in short, the Human Factors discipline—must be translated into meaningful design practices. Concepts like ease of operation or error-free performance must be interpretable in hardware and system terms.

Human Factors is one of the newer engineering disciplines. Perhaps because of this, engineering and human-factors specialists lack a common orientation with which their respective disciplines can communicate. The goal of the Wiley Human Factors Series is to help in the communication process by describing what behavioral principles mean for system design and by suggesting the behavioral research that must be performed to solve design problems. The premise on which the series is based and on which each book is written is that Human Factors has utility only to the degree that it supports engineering development; hence the Series emphasizes the *practical application* to design of human-factors concepts.

Because of the many talents on which Human Factors depends for its implementation (design and systems engineering, industrial and experimental psychology, anthropology, physiology, and operations research, to name only a few), the Series is directed to as wide an audience as possible. Each book is intended to illustrate the usefulness of Human Factors to some significant aspect of system development, such as human factors in design or testing or simulation. Although cookbook answers are not provided, it is hoped that this pragmatic approach will enable the many specialists concerned with problems of equipment design to solve these problems more efficiently.

DAVID MEISTER
*Series Editor*

# CONTENTS

I. INTRODUCTION TO TELEOPERATORS   1
   Recent History   4

II. TELEOPERATOR APPLICATIONS   9
   What Makes an Environment Hostile?   10
   Aerospace Applications   11
   Undersea Applications   19
   Nuclear Industry Applications   23
   Terrestrial Transportation   29
   Artificial Limbs   31
   Industrial Applications   32
   Public Service Applications   34
   From Puppets to Servants   35

III. SUBSYSTEMS AND THEIR INTEGRATION   37
   Subsystem Interfaces   40

IV. TELEOPERATOR DESIGN PRINCIPLES   44
   The Communications Subsystem   48
   The Computer Subsystem   57
   The Propulsion Subsystem   59
   The Power Subsystem   62
   The Attitude-Control Subsystem   69

The Environment-Control Subsystem 71
The Structure Subsystem 72

## V. THE CONTROL SUBSYSTEM 73

Makeup of the Control Subsystem 74
Man as an Element in the Control Subsystem 75
Some Special Teleoperator Control Problems 78
Performance Factors 80
Control Theory 83
The Man-Machine Interface 100
Bridging the Interface 108
Teleoperator Controls 110

## VI. THE SENSOR SUBSYSTEM 145

Direct-Vision Situations 148
Viewing with Mirrors and Fiberscopes 152
Remote Television 153
Acoustic Sensors 158
Touch Sensors 159
Displays 162

## VII. THE ACTUATOR SUBSYSTEM 171

Actuator Design Principles 173
All-Mechanical Actuator Subsystems 182
Hydraulic Teleoperators 190
Electrical Teleoperators 206
Advanced Actuator Concepts 214
Terminal Devices 215

## VIII. CONCLUSIONS AND FORECAST 219

BIBLIOGRAPHY 221

INDEX 249

# Human Factors Applications in Teleoperator Design and Operation

# I
# INTRODUCTION TO TELEOPERATORS

Early in the nineteenth century, Napoleon sat across a chessboard from a ferocious-looking automaton swathed in the robes of a Turk. Napoleon moved his chessmen into battle; the Turk did the same. Then, when Napoleon blundered three times in succession, the audacious machine swept the board clean with an iron hand.

The chess-playing Turk was constructed by Baron Von Kempelen; it took on all comers until Edgar Allen Poe deduced that beneath the Turk's chess table there was a midget chess expert who manipulated the various controls that gave "life" to the machine. Those were the innocent times when man believed that he could build anything—not the least of which was a chess-playing robot.

Now that man must work in outer space, the ocean depths, and other hazardous environments, he is building machines that recall Von Kempelen's intricate "automaton." These machines perform as appendages of man, particularly his arms, hands, and legs. Radio links, copper wires, and steel cables replace nerve fibers and muscle tendons. We shall call these man-machine systems "teleoperators," whether they are the tongs used by the old-fashioned grocer to retrieve a cereal box from the top shelf or the mechanical hand that may repair some future nuclear-powered space vehicle. The basic concept is portrayed in Fig. 1-1, where man's bodily dexterity is shown communicated across a barrier to mechanical actuators that can operate under loads too great for an unaided man, or in an environment too hostile or too far away for him to conquer in person. A teleoperator augments a normal man, or, in the case of prosthetics, helps a handicapped man become more nearly normal.

NASA is concerned with the development of teleoperators because many astronautical targets are so far away that they must be explored by proxy. Yet the amplification and extension of man via the teleoperator concept transcends space exploration. A survey of this fascinating tech-

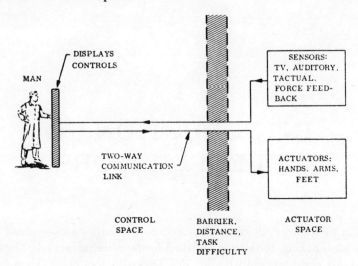

**Figure 1.1** Generalized schematic of a teleoperator incorporating dexterous actuators in the actuator space. The "barrier" between the control and operating spaces may result from distance, a hostile environment, or the sheer physical magnitude (weight, e.g.) of the task to be done.

nology must also embrace many advances made in the nuclear industry, in undersea exploration, in medicine, and in the engineering of "man amplifiers."

A teleoperator is a *general purpose, dexterous, cybernetic machine.* These adjectives separate teleoperators from other machines. The adjective "cybernetic" excludes all preprogrammed machinery, such as timer-controlled ovens, record-changing phonographs, and much of the machinery on automatic production lines. A teleoperator, in contrast, *always* has man in the control loop. The other adjectives—"dexterous" and "general purpose"—sharpen the focus further. These semantic sieves trap human-controlled, but undexterous, machines such as remotely controlled aircraft and telephone switching circuits. The man-machine systems that fall through our sieves allow man to do the following:

—Pick up and examine samples of the lunar surface while remaining on Earth.
—Repair an underwater oil pipeline from a surface ship.
—Manipulate radioactive nuclear fuel elements in a hot cell.
—Regain dexterity with an artificial limb (the prosthetics concept).
—Lift a ton-sized load (the man-amplifier concept).

The prefix "tele" in teleoperator describes the ability of this class of man-machine systems to project man's innate dexterity not only across distance but through *physical barriers as well.*

When an area of technology with latent commercial potential approaches that point where exponential growth appears imminent, engineers invariably become word testers. Because no unified discipline welds the technical innovators together, synonyms and overlapping words proliferate. The following glossary should dispel some of the confusion:

> *Telepuppet.* A word coined in the 1950's by Fred L. Whipple, now director of the Smithsonian Astrophysical Obeservatory, to describe his concept of how sophisticated machines could take the place of man on spacecraft. The word has not become popular, presumably because "puppet" implies toys and entertainment rather than science and engineering.
>
> *Telechirics.* John W. Clark synthesized this word from Greek roots while at Battelle Memorial Institute in the early 1960's (Clark, 1963). Literally, telechirics means "remote fingers." It is descriptive, but unfortunately excludes walking machines and man amplifiers.
>
> *Telefactor.* The idea of making or doing something at a distance is intrinsic in this word conceived by William E. Bradley, at the Institute for Defense Analyses (Bradley, 1966). It is semantically sound, but many people do not immediately recall that "factor" implies doing or making as well as algebra.
>
> *Cybernetic anthropomorphic mechanism (CAM for short).* Ralph S. Mosher, at General Electric, has often used this term in his papers on walking machines (Mosher, 1964), but it excludes many non-anthropomorphic mechanisms included in this survey. Mosher now refers to the field as *mechanism cybernetics,* a term that omits only the desired attributes of dexterity and versatility.
>
> *Master-slave.* Originated by Ray C. Goertz at the AEC's Argonne National Laboratory in the late 1940's, this term is generally applied only to the common mechanical and electronic manipulators that have long been used in hot cells (Goertz, 1964).

The terms "manipulator" and "remote control" are also often associated with the telemechanism field. The first term is too narrow a concept, since it excludes walking machines and exoskeletons. "Remote control" is too broad because it includes everything man does at a distance, even to changing a TV channel from his armchair.

A compact, accurate synonym for general purpose, dexterous cybernetic machines may evolve as the field matures. Meanwhile, "teleoperator" will serve in this book.

## 4   Introduction to Teleoperators

Figure 1.2 rounds out the picture of the teleoperator by portraying its full set of subsystems. Four of the nine subsystems deal directly with machine augmentation of man:

—*The actuator subsystem* that carries out the manipulations and other dexterous activities ordered by the human operator. The actuators may be stronger, more dexterous, and faster than the operator.
—*The sensor subsystem* that permits the operator to see, feel, hear, and otherwise sense what the actuators are doing in the actuator space and what their environment is.
—*The control subsystem*, which includes the human operator, analyses information fed back by the sensors in the actuator space and compares this with the operational objectives. The result is a series of commands to the actuator subsystem.
—*The communication subsystem* is the information hub of the teleoperator. It transmits commands and feedback among the various subsystems.

The supporting roles of the other five subsystems shown in Fig. 1.2 are apparent from their names. Chapter 3 will elaborate on the parts played by the nine subsystems and describe how they act in concert to carry out man's directives.

While the system diagram may seem somewhat involved, it is sufficiently general to include simple tongs for handling radioactive samples and extremely complex systems.

### RECENT HISTORY

The chess-playing Turk was preceded by the marvelous automatons of the Jaquet-Droz father-son team in the late 1700's (Porges, 1957). Controlled by grooved, rotating disks, the Jaquet-Droz automatons could play music and write out compositions; one in particular, "The Draughtsman," astounded King George III and Queen Charlotte by sketching them on the spot—or so it seemed. (Such a machine would be called preprogrammed today.) A "Steam Man," built by a Canadian, Professor George Moore, in the 1890's, was powered by a half-horsepower, high-speed steam engine; this primitive walking machine could puff along pulling light loads behind it. The Westinghouse automatons exhibited at the New York World's Fair in 1939, "Elektro" and "Sparko," could walk, talk, and distinguish colors. The word "robot" means "worker" in Czech and gained popularity from Karel Capek's 1923 play "R.U.R." (for "Rossum's Universal Robots"). Today a robot is generally considered to be

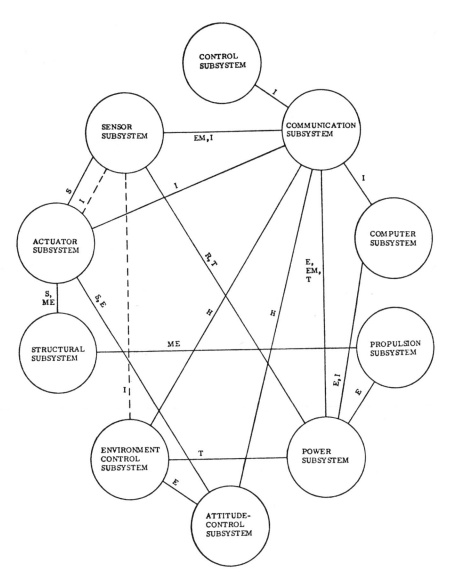

**Figure 1.2** Interface diagram for teleoperators. Some of the most important interface forces between subsystems are indicated by the following code: S = spatial, E = electrical, EM = electromagnetic, R = radiative, ME = mechanical, T = thermal, I = information. (See Fig. 3.1 for examples.) A dotted connecting line indicates a local control loop that bypasses the control subsystem, such as a thermostat temperature control.

an automaton made in the shape of a man. Robots are usually preprogrammed or, in science fiction particularly, self-adapting and intelligent, not requiring and even disdaining help from humans. In constast to robots, man is always intimately in the loop in the teleoperators discussed in this book.

Taking the historical road labeled "teleoperators," let us pass over the early and well-documented developments of television, cybernetics a lá Norbert Wiener, radio control, and the supporting technology of prosthetics, and begin with master-slave manipulators built under the impetus of the atomic energy program. These were the first really sophisticated machines to project man's manipulative capability into a hazardous environment.

The chronology runs like this:

—*1947*. Mechanically and electrically connected unilateral* manipulators were developed at the AEC's Argonne National Laboratory (ANL).

—*1948*. Ray Goertz and his coworkers at ANL developed the Model-1 bilateral mechanical master-slave manipulator (Goertz, 1964).

—*1948*. John Payne built a mechanical master-slave manipulator at General Electric (Anon., 1948), and many AEC installations subsequently acquired a great variety of mechanical manipulators.

—*1948*. General Mills produced the Model-A unilateral manipulator in which the arms and hands were driven by switch-controlled motors rather than by direct mechanical or electrical linkage to the operator (as in the true master-slave). The Model-A became a "workhorse" of the nuclear industry in tasks requiring more strength and working volume than possible with master-slaves.

—*1950*. ANL experimented with master-slaves coupled with stereo TV.

—*1954*. Development of the Argonne Model-8 mechanical master-slave manipulator was completed. This manipulator is still predominant in the atomic energy industry and is manufactured commercially.

—*1954*. Ray Goertz built an electric master-slave manipulator incorporating servos and force reflection (sense of touch or "feel") (Goertz, 1964). The master-slave position control of the manipulator arms and hands plus force reflection made this the first bilateral electric manipulator.

—*1954*. The GPR (General Purpose Robot) was built at the AEC's

___
*"Unilateral" means that there is no kinesthetic or force feedback as there is in a "bilateral" system. See Table 3.1 for definitions of the various kinds of teleoperators.

Savannah River Plant. This was the first, general purpose manipulator-equipped vehicle.

—*1957.* Professor Joseph E. Shigley, at the University of Michigan, built a primitive walking machine for the U.S. Army (Shigley, 1960). Although many walking machines were built earlier, Shigley's inaugurated the present-day Army program in "off-road" locomotion.

—*1958.* First mobile manipulator with TV was built at ANL. This teleoperator was called a "slave robot."

—*1958.* Ralph S. Mosher and coworkers at General Electric built the Handyman electrohydraulic manipulator incorporating force feedback, articulated fingers, and an exoskeletal control harness. This equipment was built for the joint AEC-USAF Aircraft Nuclear Propulsion Program (Mosher, 1960).

—*1958.* William E. Bradley, Steven Moulton, and associates at Philco Corporation developed a head-mounted miniature TV set that ennabled an operator to project himself visually into the operating space.

—*1961.* The first manipulator was fitted to a manned deep-sea submersible when a General Mills Model 150 manipulator was installed on the *Trieste* (Hunley, 1965).

—*1963.* The U.S. Navy began deep-submergence projects, including the development of underwater manipulators.

—*1963.* R. A. Morrison and associates at Space-General Corporation constructed a lunar walking vehicle. This machine was later converted into a "walking wheelchair" for handicapped children (see Chapter 2).

—*1964.* Neil J. Mizen and coworkers at Cornell Aeronautical Laboratory reported on the construction of a "wearable, full-scale, exoskeletal structure." The Cornell exoskeleton was not powered (Mizen, 1964).

—*1965.* Ray Goertz and his associates at ANL combined the ANL Model E4 electrical master-slave manipulator with a head-controlled TV camera and receiver (Goertz, 1964).

—*1966.* ANL combined the Model E3 electric master-slave with the Mark TV2, head-controlled TV, which added translational motion to the viewing system.

—*1966.* Case Institute of Technology, working under a NASA grant, demonstrated a computer-controlled manipulator that can perform preprogrammed subroutines specified by the operator.

—*1967.* Variety of tactile sensors demonstrated by J. C. Bliss at Stanford Research Institute.

*—1968.* Head-controlled television with split foveal-peripheral visual display developed for moving vehicles by J. Chatten at Control Data.

*—1968.* J. Allen and A. Karchak, at Rancho Los Amigos Hospital, construct position-controlled anthropomorphic manipulators.

This chronology gives little hint of the imminent and intimate man-machine partnership that many believe essential to the large-scale exploitation of space and the oceans. Many of the most important developments listed were made under the aegis of the Atomic Energy Commission. Further developments are likely in many fields, as man learns to let machines do the "dirty work" while he thinks more.

# II
# TELEOPERATOR APPLICATIONS

Since 1948, some 3,000 manipulator arms have been built in the United States. More than 80 percent were shipped to atomic energy installations where visitors can see them lined up in precision formation like well-disciplined metallic soldiers (Fig. 2.1). These long banks of master-slaves are only the advance guard of an army of man-machine systems now being assembled to serve man in a variety of ways.

Some applications of teleoperators, such as the lifting and manipulation of a two-ton crate, or the tactile inspection of a long, narrow, serpentine passage, result from the human body's limited strength, fixed size, and restricted articulation. Most teleoperators, however, are applied in so-called "hostile" environments from which man is excluded by high temperatures, nuclear radiation, or the crushing pressures of sea water. Asbestos suits, diving gear, and space suits let man temporarily enter these dangerous realms, but his stay is usually brief and expensive.

Economy often determines the choice between man and teleoperator. It is likely to be cheaper, for example, to send a diver down to make pipe connections in shallow off-shore oil fields than to develop a teleoperator to do it. Below 100 fathoms, however, divers are encumbered by heavy suits and cannot stay down long. Deep diving is so costly that teleoperators may dominate the deep oil fields far out on the continental shelves.

The bulk of today's operational teleoperators are those unilateral and master-slave manipulators installed in hot cells to handle radioactive materials. General purpose manipulators are used because they are cheaper than a multitude of special purposes machines. Manipulators enable personnel and facilities to operate more efficiently without waiting for radioactive materials to decay to levels at which they can be handled by men directly. Ray Goertz, who pioneered the development

## Teleoperator Applications

**Figure 2.1** Typical bank of mechanical master-slaves at AEC's CANEL facility, at Middletown, Conn. (Courtesy of Pratt & Whitney Aircraft.)

of the master-slave manipulator at Argonne National Laboratory, estimates that the introduction of the master-slave is saving the nuclear industry well over 15 million dollars per year in operating costs and roughly 15 million dollars additional per year on special equipment and facility costs. Teleoperators will probably succeed wherever they can muster similar, convincing economic arguments.

In summary, four considerations help determine when a teleoperator will augment man:

1. Man's absolute physical limitations in matters of strength, endurance, size, and bodily construction.
2. Human welfare and safety.
3. Dollars-and-cents considerations.
4. Farther in the future, aesthetics. This implies that some tasks are too "dirty" or demeaning.

## WHAT MAKES AN ENVIRONMENT HOSTILE?

In South America, some Indians live nearly naked in the frigid winters of Tierra del Fuego; others live high on the rarefied peaks on the Andes, aided by abnormally large lung capacities. Despite man's astonishing ability to adapt to the Earth's varied climes, he often covets air conditioners or furnaces. When he enters environments more hostile than those found on the earth, he attempts to encapsulate and carry a comfortable environment along with him. The harsher the environment that a "canned man" invades, the more expensive and inconvient the can.

One way to show how teleoperators can aid man is to list "hostile" forces and factors that man cannot handle conveniently alone. Table 2.1 does this and, at the same time, suggests rather intimate man-machine symbiosis. Quite obviously, this man-machine intimacy derives not only from the strength, and hardiness of teleoperators, but also from their hopefully superior senses, reaction times, and abilities to handle (with the aid of computers) complex tasks. Teleoperators amplify and extend the normal man and enhance the capabilities of the physically handicapped.

Although designed to replace men in hazardous environments, teleoperators often are far from invulnerable themselves. For example: teleoperators, if they are to emulate man, must have articulated limbs and the joints must be kept free from seawater-borne silt if they are used on submersibles; in hot cells they must be lubricated with grease that does not degrade under the influence of nuclear radiation. If a teleoperator is to operate in a high temperature, the electronics subsystems in particular must be cooled to preclude degradation. Teleoperators in areas of radioactive dust or dangerous biological agents must not permit these contaminants to leak through the barrier separating man from the hazardous environment. These are only a few of the design constraints dictated by the application. A hostile environment is also hostile to machines, but less so.

Table 2.2 summarizes present and proposed applications of teleoperators to various industries.

## AEROSPACE APPLICATIONS

Astronauts and special purpose remote-control machines perform today's manipulative tasks in outer space. An astronaut is vulnerable, expensive, and non-expendable. Special-purpose machines, such as the Surveyor surface sampler, which preceded man to the Moon and performed the first crude manipulatory experiments with lunar soil and rocks, are useful but neither particularly dexterous nor versatile.* Dexterous, rugged, general purpose teleoperators can be further developed to aid or replace men and special purpose machines. Limited commu-

*There are no well-defined "thresholds" of dexterity or versatility that separate teleoperators from simple tools.

Table 2.1 Environmental Properties Affecting Teleoperator Selection

| "Hostile" environmental properties[a] | Typical environments | Current solutions |
|---|---|---|
| High temperature | Metal-treating plants, fires | Heat shields, asbestos suits, gloves |
| Low temperature | Outer space, arctic regions | Space suits, insulated clothing |
| High pressure | Undersea | Armored diving suits and bells, teleoperators |
| Low pressure | Outer space, vacuum chambers | Space suits, teleoperators |
| Toxic atmosphere | Mining, warfare, many industrial processes | Suits and masks |
| Nuclear radiation | Hot cells, nuclear accidents, radiotherapy, space | Portable shields, teleoperators |
| Acoustic | Airfields, launch pads, rockets | Ear covers, absorbers |
| High acceleration, jostling | Aircraft, rockets, spacecraft landings | Special suits and harnesses |
| Zero gravity | Spacecraft | Artificial gravity |
| Sickening or disorienting motion | Spacecraft and other types of transportation | Drugs, stabilizers |
| Projectiles | Space (meteoroids), mining, blasting | Armor, shields |
| Biological | Warfare, laboratories | Biological barriers, quarantine, immunization, masks, gloves |
| High forces, heavy weights | Everywhere | Special-purpose machine (tools) teleoperators, prostheses |
| Complexity (too many objects, tasks, targets) | Can occur anywhere | Computer help. More than one operator |
| Endurance (one of the oldest reasons for introducing machines) | Can occur anywhere | Various special machines (but not teleoperators which always have man in the loop), shifts of men |

Table 2.1  Environmental Properties Affecting Teleoperator Selection—Concluded

| "Hostile" environmental properties[a] | Typical environments | Current solutions |
|---|---|---|
| Speed of targets | Can occur anywhere | Computer help, various special machines |
| Precision movement | Electronics construction, biochemical industry, surgery | Micromanipulators |
| Small and/or serpentine task apertures | Can occur anywhere | Various special machines, and tools, teleoperators |
| Entertainment value | TV, stage, fairs, parades | Puppets, Disney creations |
| Sensory blackout (loss of visual, acoustic, and/or tactile contact) | Undersea, space, polar regions | TV, microphones, sonars, tactile probes—all are teleoperator subsystems |

[a] Several of these properties may be present simultaneously in a hostile environment.

nication bandwidth has slowed the introduction of teleoperators but the situation is improving. William E. Bradley has suggested some intriguing advantages and disadvantages of teleoperators beyond those already suggested; viz.: hardiness, endurance, relative invulnerability, etc. (Bradley, 1966).

— A teleoperator has total recall because it is possible to record back on Earth all the machine does and sees.
— The visual scenes transmitted by the teleoperator can be easily retransmitted over worldwide television, giving viewers the sense of being direct participants in extraterrestrial feats.
— A true automaton with self-adaptive capabilities does not require the interplanetary communication capacity of the teleoperator, but the teleoperator with man in the loop should be more versatile and self-maintaining.
— At lunar and planetary distances, teleoperators suffer time-delay problems such that the Earth-based operator can not see the results of his actions for several seconds or even minutes. This factor may severely limit the employment of teleoperators on distant missions.

Table 2.2  Summary of Teleoperator Applications

| Industry | Present and/or proposed application |
|---|---|
| Aerospace | Rarely used in aircraft at present. Occasionally used in vacuum chambers and in handling propellants and explosives. Proposed for spacecraft assembly and maintenance and for exploration of Moon and planets. Man amplifiers proposed for high-g operation and cumbersome space suits. The Surveyor surface sampler was a crude teleoperator. Suggested for inspection and repair of satellites by remote control. |
| Undersea | Manipulators are installed on nearly all new research and rescue submersibles. Also used for weapons recovery, ship salvage, and rescue. Used to limited extent in off-shore oil field operations and the repair and maintenance of undersea laboratories and military devices. |
| Nuclear | Used in hot-cell operations with radiochemicals, fuel fabrication and reprocessing, inspection of radioactive equipment, and production of radioisotopes. Used in emergency situations for inspection, rescue, cleanup, and decontamination. Used for inspection and disassembly of nuclear reactors. Accelerator repair and maintenance. |
| Terrestrial transportation and material handling | Walking machines under development for off-road military transportation. Man amplifiers being designed to augment lifting and carrying capabilities of individual soldiers. Suggested for minefield clearing, lumber industry, warehousing, etc. |
| Medical | Prosthetic and orthotic devices used for many years. Walking wheelchairs and man-amplifiers proposed for handicapped. Teleoperators suggested for remote surgery and microsurgery. |
| Chemical and biological | Limited use in handling toxic materials, propellants, and explosives. Proposed for handling dangerous biological agents in the laboratory. |
| Metal processing, handling, and fabrication | Long used in forging plants and for handling large, hot metal pieces. |
| Electronics | Proposed for super-clean rooms and operations in toxic atmospheres. |
| Construction and mining | Proposed for explosive environments. |
| Public services | Proposed for fire-fighting and for rescue and cleanup in hazardous environments, such as gasoline, chlorine, and radioisotope spills. |
| Entertainment | Long used where the human operator wishes to remain concealed as in puppet shows, mechanical men, and animated creatures in à la Disney. |

## Testing, Stimulation, and Sterilization

Radars and other avionic gear must be tested in chambers that simulate high altitudes or space conditions. Repressurization of a big chamber just to flick a console switch is obviously inefficient. A simple manipulator piercing the side of such a chamber may solve the problem. Typical of this application is the AMF Mini-Manip installed at the Norden Division of United Aircraft Corporation in Norwalk, Conn. The impetus for using a teleoperator here is purely economic; chamber time and engineer's time are too expensive to waste in avoidable chamber repressurizations.

Manipulators may also find application in *very large* environmental test chambers in which full-scale manned space vehicles are tested. In 1963, the General Electric Company completed a study for the Air Force's Arnold Engineering Development Center a Tullahoma, Tenn. (Olewinski, 1963). In the large chamber studied (220 feet in diameter) manipulators were proposed for such routine tasks as the placement and adjustment of radiation sources and simple "switch-throwing" operations like those described for the Norden chamber. The manipulator would thus relieve astronauts of tasks unrelated to the vehicle tests. A more dramatic task proposed or manipulators was the rescue of astronauts should serious injuries or life-support equipment failure occur. The GE study suggested use of both long, boom-mounted manipulators and small vehicles with manipulators, similar to those built for large hot cells. Rapid chamber repressurization was not considered an acceptable solution to the rescue problem in this study.

Several persons have suggested building teleoperators in man-like form to replace aircraft and spacecraft test pilots. A teleoperator could manipulate the vehicle's controls without risking human life but the concept is practicable *only* when the equipment being tested will eventually have a human operator at the controls; otherwise ordinary remote control could be used.

An airplane out of control may produce such violent accelerations (jostling and high-g forces) that its pilot is incapable of moving the controls or even operating an ejection mechanism (Loudon, 1964). A powered, partial exoskeleton can come to his rescue by allowing him to move an arm voluntarily to the ejection control switch. General Electric has suggested use of a servo restraint harness system to help a pilot operate aircraft controls under high-g conditions.

Humans have dexterous hands but these same hands carry microorganisms and various kinds of "dirt" that can and do contaminate spacecraft parts during construction. Even a carefully masked and clothed human may carry some aura of microbes and "dirt." Here lies one of the major problems in the aerospace and many other industries: clean assembly (Lorsch, 1966). Why not master-slave use manipulators for parts assembly? This is the radioactive hot-cell problem in reverse: i.e., keeping contamination out instead of in. This is still virgin territory for teleoperators.

**Satellite and Deep-Space Operations**

The spectrum of tasks proposed for teleoperators in orbit and deep space is so broad that a list is in order to provide perspective.

—Satellite inspection to identify status, malfunctions, or fix its purpose and country of origin.
—Satellite capture and de-spin.
—Satellite maintenance and repair, particularly space vehicles incorporating nuclear power plants or propulsion systems. (General Electric, 1969).
—Satellite turn-off, supposing its "killer timer" has malfunctioned.
—Attachment of deorbiting rockets.
—Satellite destruction or disarming of military satellites.
—Satellite assembly and test. The erection of large space antenna arrays has been suggested as a promising application for teleoperators (Bradley, 1966).
—Removal and/or replacement of experiments and samples (such as coupons to measure micrometeoroid damage).
—Satellite experiment modification, rearrangement, or adjustment; viz., changing filters and photographic plates in an orbiting telescope.
—Aiding spacecraft docking.
—Propellant and cargo transfer, particularly if the cargo is hazardous.
—Astronaut rescue, which might involve satellite de-spin, forcible entry, and transfer of a man to a rescue vehicle.
—Exoskeletons to improve an astronaut's mobility and dexterity.

Many of the above needs might arise during the same mission. Since it would be inefficient to build a different machine for each task, one of the selling points of space teleoperators is their versatility and generality. As a consequence, most studies of teleoperators for space and deep-sea work have focused on general purpose vehicles bristling with manipulator arms. Space vehicles carrying teleoperators bear such fanciful names and acronyms as Remora, Humpty Dumpty, Man Friday, and MEMU.

A major NASA study effort was completed in 1966 when Ling-Temco-Vought (LTV) and Argonne National Laboratory (ANL) investigated a Maneuvering Platform (MWP) and a "Space Taxi" with attached manipulators for Marshall Space Flight Center (Ling-Temco-Vought, 1966). This was a study of the utility of a manned maneuvering space capsule on such potential missions as the Manned Orbiting Laboratory (MOL), the Apollo Applications Program (AAP), the Manned Orbiting Research Laboratory (MORL), and the Orbiting Launch Facility (OLF). Later chapters will cover the MWP and Space Taxi concepts in more detail.

A second important manipulator study was concluded by General Electric in 1969, also under NASA contract. (General Electric, 1969) Taking a different approach than the LTV-ANL work, the GE groundrules kept man on the ground. The manipulator operator saw his work via an orbit-to-Earth communication link instead of directly from a man-rated orbital vehicle. NASA's purpose in the GE study was to determine the feasibility of remote repair, maintenance, and resupply of large unmanned satellites, such as the Orbiting Astronomical Observatory (OAO) and Orbiting Solar Observatory (OSO). GE concluded that such expensive satellites could have their useful lifetimes extended by manipulator-carrying spacecraft for costs far less than those of brandnew satellites.

Many orbital teleoperator concepts look like extra-terrestrial bugs. Generally, man is enclosed in a spherical or cylindrical capsule under shirtsleeve conditions. He controls special arms that grasp the target and firmly anchor the space capsule to it. Other controls move the working arms outside the capsule. Because space is precious on spacecraft (and on small submersibles), the master side of space manipulators is usually miniaturized. Figure 2.2 shows one conceptual design for an orbital capsule. The GE concept of a satellite repair and maintenance vehicle is portrayed in Fig. 2.3.

**Planetary Operations**

Exploration of the Moon and other planets thus far has fallen to unmanned, special-purpose remote-control machines, such as Rangers, Mariners, and Surveyors. Remote control on such space vehicles is confined to switch throwing and the initiation of programmed sequences; viz., Mariner's planetary-scan platform. With teleoperator arms and hands, an Earth-bound operator could direct manipulations impossible with special-purpose remote-control systems. With teleoperators on a large, unmanned, planetary lander one might:

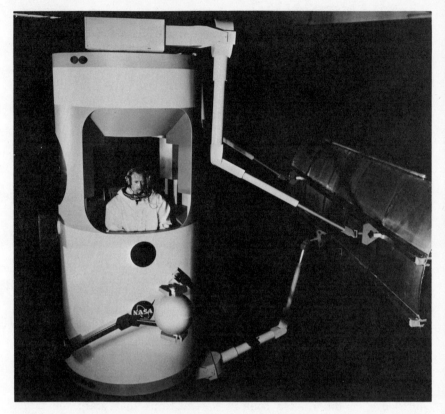

**Figure 2.2** Mockup of the Space Taxi designed by Ling-Temco-Vought and ANL for NASA's Marshall Space Flight Center for orbital repair and maintenance work. A complete Space Taxi would have three docking arms and two "working" arms.

—Vary, adjust, and modify experiment layout.
—Maintain and repair equipment.
—Collect and handle samples with great flexibility.

An automated, unmanned laboratory on Mars or Venus, for example, might benefit from Earth-controlled manipulators. Such a laboratory would then be analogous to the undersea Benthic Laboratory conceived by Scripps Oceanographic Laboratory and discussed later in this chapter.

Two major disadvantages of employing teleoperators to study the Moon and planets are (1) the time-delay factor, and (2) the very wide bandwidths needed to handle television and control signals for a many-jointed teleoperator. The precise point at which teleoperators may be-

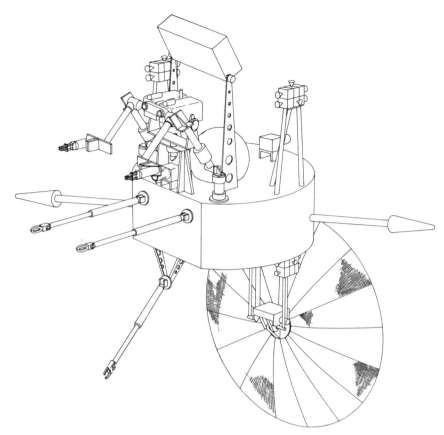

**Figure 2.3** A manipulator-carrying, unmanned spacecraft proposed by General Electric for the repair, maintenance, and resupply of scientific and applications satellites. (General Electric, 1969.)

come cheaper and more effective than limited purpose, remote-control exploratory machines like the Surveyors is not known.

## UNDERSEA APPLICATIONS

Almost all of the ocean floor is at least two miles deep. Even on the shallow continental shelves, divers rarely work below 100 fathoms. The military threat of hostile vehicles and installations makes it imperative that we know how to work under water. Substantial petroleum reserves under the deeper portions of the continental shelves have given commercial impetus to undersea technology. Undersea manipulators have already recovered debris from the sunken *Thresher* and an errant H-bomb, though with great difficulty in each instance.

Although many operational problems of inner and outer space are similar (viz., the necessity of firmly anchoring the teleoperator vehicle to the target), the environments have radically different effects on teleoperator design. The undersea teleoperator is surrounded by a good heat sink, but one that is extremely corrosive and laden with silt and biological agents. The tremendous pressures at great depths preclude the common mechanical master-slave linkages between the control and actuator spaces. The sensor problem is also different. Instead of the bright sunlight of orbital space, there may be such darkness that an operator cannot see a manipulator hand which is only a few feet in front of his viewport.

Both in outer space and under the sea men may have to identify, build, maintain, repair, recover, or destroy some object. These activities require cleaning, bolting, cutting, welding, replacing parts, etc.—just the things men's hands do to terrestrial, dry-land equipment. In the oceans the missions may be for (1) scientific research, (2) commercial operations, or (3) military operations.

## Undersea Scientific Research

The small, manipulator-equipped submersible is common to all three mission classes. In early bathysphere descents, scientists were passive observers. Even the simplest manipulators widen research horizons considerably, as Fig. 2.4 demonstrates with Cousteau's Diving Saucer and its manipulator-captured nimble prize. A more advanced submersible concept is the North American Aviation, Inc., *Beaver*. Other manipulator-equipped submersibles include the *DOWB, Alvin II, AUTEC I, Trieste II, Deepstar, Sea Cliff,* and *Turtle* (Hunley, 1965). These are general purpose utility craft capable of manipulating objects outside of the protective hull sheltering the human operator(s).

Selective sampling is much more effective than hit-or-miss dredging from surface ships. Submarine geology will profit immensely as manipulators bring back rocks, nodules, and deepsea ooze samples. Shells, plants, and the more sluggish forms of marine life are targets, too. Manipulators can also set up, maintain, and repair such undersea scientific equipment as gravimeters, current meters, seismometers, corers, and penetrometers. For archeologists, submersibles such as *Asherah*, fitted with teleoperators, can retrieve artifacts and help with underwater excavations.

Victor C. Anderson, of Scripps Institution of Oceanography (University of California), has described the Marine Physical Laboratory's Benthic

Figure 2.4 The hydraulically actuated manipulator (called a "clamshell grab") on the Diving Saucer SP–300 holding a spider crab. (Courtesy of Westinghouse Electric Corp.)

Laboratory: an unmanned, self-repairing, self-maintaining, ocean-floor capsule fitted with manipulators (Anderson, 1964). The Benthic Laboratory is built according to a modular philosophy that enables the manipulator located inside to replace electronics components and modify experimental setups. The "autonomous" nature of the Benthic Laboratory has much in common with self-contained hot cells that operate sealed up for years. Such a capability is ideal for *in situ* scientific experiments both on the ocean floor and on distant planets. One of the first uses of the Benthic Laboratory will be to support a "sensor field" of current meters on the floor of Scripps Canyon off California.

Another teleoperator concept is the bottom crawler equipped with manipulators, lights, and television. Except for the trailing power and control cable, the Scripps remote-controlled, underwater manipulator (RUM) vehicle might be considered the "wet" analog of NASA's lunar and planet crawlers. The missions of RUM-type vehicles would be similar to those of the small submersibles. The sea bottom, however, is treacherous territory, and "hovering" submersibles have proven to be more versatile and mobile.

Walking machines are of questionable merit on the sea floor because of the precarious footing. A powered exoskeleton, however, might materially aid a heavily armored diver by permitting him to work longer and carry out tasks requiring more than human strength.

## Commercial Underwater Operations

In 1966 more than 1800 offshore oil wells were operating from surface platforms in an average depth of 200 feet of water. Divers currently perform the many underwater tasks necessary to bring an offshore well into production. Drilling operations, however, are moving out into water so deep that divers can work in it neither effectively nor for long periods. With few exceptions, the manipulator-equipped small submersible is the instrument attractive to the interested oil companies. The same submersibles built for underwater research may help bring in petroleum from the continental shelves.

Task surveys show a wide range of jobs for teleoperators:

—Surveying and selecting drill sites.
—Preparing the drill sites.
—Observing and assisting during drill string landing.
—Replacing of blowout-preventer rams.
—Making "completions"; i.e., pipe connections.
—Replacing and patching pipe sections.
—Recovering objects dropped from drill platforms.
—Removing marine growth.
—Routing and installing pipelines.

Hughes Aircraft Company built the UNUMO and a MOBOT for trials in offshore oil fields (Hunley, 1965) before the advent of the small submersible. The UNUMO was a ship-suspended teleoperator carrying lights, attachment arms, manipulators, television camera, and a propulsion system. Its mobility and versatility were limited, however, and it was never put into operational use. Hughes also built a version of the MOBOT for the Shell Oil Company for undersea trials, but it has not had widespread use.

An intriguing commercial application of teleoperators is in salvage work—or even treasure hunting. Hunley and Houck report that the submersible *Recoverer I* has been employed in raising a 165-foot sunken fishing vessel off Cape Lookout, North Carolina (Hunley, 1965). Some representative manipulator tasks were clearing away debris and rigging, attachment of flotation containers, slinging cables, closing valves, and placing explosive charges for cutting away standing rigging.

A much-advertised commercial aspect of deep-sea exploration has been mining of the manganese nodules that pave many sections of the ocean floors. Picking up these nodules one by one with manipulators would not be economical, but teleoperators could certainly be employed in surveying and sampling nodule fields for eventual mining.

In all commercial applications, the indifference of teleoperators to time, fatigue, and the hostile properties of the deep-sea environment is of fundamental economic importance. Keeping ships at sea and divers on the botton are costly operations. The advantages of around-the-clock teleoperators are obvious.

**Military Underwater Operations**

Small unmanned, sea-floor stations perform the same functions as navigation and reconnaissance satellites. Like their space cousins, they must be installed, maintained, and repaired, and such tasks may warrant further devolpment of teleoperators.

The *Thresher* catastrophe in 1963 and the H-bomb recovery off Spain in 1966 reinforced the status of teleoperators in undersea military activities. The H-bomb was recovered by a teleoperator called CURV (Cable-controlled Underwater Research Vehicle), which the U.S. Naval Ordnance Test Station had previously employed for operations such as torpedo recovery (Heller, 1966). CURV is equipped with high resolution sonar, television camera, three screws for propulsion, and a rather crude claw for grasping objects.

The *Thresher* incident spawned a series of small submersibles, similar to those employed in scientific and commercial activities, but for personnel rescue. The first submersible in this series to be built was the *DSRV-1* (Deep Sea Rescue Vehicle) and Lockheed Missiles & Space Company was the prime contractor. The *DSRV-1* is a small, nuclear-powered submarine, transportable in a C-141 and piggyback on a submarine. Manipulator hands will clear away debris, cut cables, and help the *DSRV-1* mate ("dock" in space lingo) with a stricken submarine and begin rescue.

## NUCLEAR INDUSTRY APPLICATIONS

The plutonium production plants of the Manhattan Project produced the first large quantities of radioisotopes during the early 1940's. The glove boxes previously employed in handling toxic materials proved completely inadequate in the highly radioactive "cells" at Hanford, Oak Ridge, Los Alamos, and other AEC installations. Long tongs alleviated the problem somewhat, particularly those with ball joints that could work through hot-cell walls, and crane operators became very adept at "manipulating" hot cargo with hooks and special attachments. Nevertheless, more dexterity was desperately needed in radiochemistry and nuclear fuel operations, and the nuclear industry is now the largest user of teleoperators.

During the nuclear weapons program, chemists faced the job of untangling hundreds of radioactive fission products found in spent nuclear fuel. They also had to develop chemical processes for extracting the plutonium from irradiated uranium fuel slugs. Once the plutonium was recovered, ways had to be found to dispose of liquid wastes—some so radioactive that they boiled spontaneously. After weapons tests, radioactive fallout had to be monitored and analyzed. The upshot of these requirements was that chemical and physical manipulations with hot materials ran the full spectrum of tasks found in conventional chemical laboratories: i.e., pouring, stirring, powdering of samples, loading furnaces, titrating, collecting evolved gases, and similar deft handling jobs. To carry out such operations through several feet of concrete and lead, chemists have learned to work with master-slave manipulators.

Hot laboratories offer so many examples of teleoperator applications that it is impractical to list them all. The bank of manipulators in (Fig. 2.1) is typical of the hundreds of hot laboratories around the world. Many glove boxes and specialized remote-control devices are still used, too. Remotized saws, drills, balances, and grinders carry out much of the repetitive work, while the manipulators are reserved for nonroutine operations, such as setting up a lathe and handling samples.

## Fuel Fabrication and Reprocessing

As nuclear power has become a big business, fuel fabrication and reprocessing have moved out of the laboratory onto the production line. The more automated the production line, the less need there is for general purpose manipulators. Nevertheless, automated equipment must be maintained and repaired; if the production or reprocessing line is very hot, manipulators will be installed for these functions. There is also a small but significant residue of tasks that cannot be automated, such as the retrieval of errant fuel pellets. Just as the most highly automated factory still employs human workers, nuclear fuel production plants will always have manipulators.

The fabrication of fuel elements from fresh uranium rarely requires more than glove-box operation because radiation levels are low. Today most reactor fuel is made without manipulators. However, as "recycle" fuel (i.e., "unburnt" uranium and plutonium from reprocessed "spent" fuel elements) enters fuel fabrication plants, glove boxes must give way to hot cells. Plutonium-240 and other radioactive constituents make recycle impossible to handle safely with glove boxes.

Such fuel-handling problems plagued the designers of the EBR–II (Experimental Breeder Reactor) Fuel Cycle Facility at the AEC's National Reactor Testing Station in Idaho (Stevenson, 1966). Hot spent fuel pins from the EBR–II had to be processed and the extracted, still-fissionable fuel refabricated into new fuel elements for reinsertion in the reactor. The circular production line begins and ends at the reactor. Long fuel assemblies pulled from the reactor enter at the left and move counterclockwise around the circle. The external metal tube is first stripped, then the enclosed fuel pins are melted and refined. After the unfissioned fuel is extracted by wet chemistry, it is fabricated into new pins. With manipulators helping at each step along the way, new fuel enters the reactor at the completion of the circle.

The EBR–II Fuel Cycle Facility was originally designed to be more highly automated than practical considerations finally permitted. For example, the fuel-pin dimensions could not be controlled with sufficient accuracy to be acceptable to all automated fuel-handling equipment on the line. In anticipation of such problems, master-slaves and specially designed, radiation-resistant unilateral manipulators had been installed and they were able to take over when automatic equipment faltered. In terms of the original design, the manipulators were partly redundant as far as fuel handling was concerned. Redundancy turned out to be good design practice, for the EBR Fuel Cycle Facility has operated successfully and continuously for more than three years without human entry.

Nuclear fuel fabrication and reprocessing generally require high-load-capacity manipulators with large working volumes. Electric unilateral manipulators are used in preference to master-slaves in most applications of this type.

### Handling Power Plants

Some of the largest teleoperators have been built for disassembling reactors destined for nuclear rockets and aircraft. During the development of these high-temperature engines, reactors are tested in a remote site and then carried to large hot cells, where they are stripped down piece by piece, fuel element by fuel element, to determine what transpired during the tests. Even after extensive cooling periods, these reactors are still radioactively hot and can be dissected only by long-reach, heavy-duty manipulators. In the nuclear rocket program, 14,000-pound, hot NERVA reactors are taken from the test stands to the E–MAD building (Engine Maintenance, Assembly, and Disassembly), in Nevada, where "rectilinear" manipulators in the Wall Mounted Handling Subsystem (WMHS) systematically disassemble them (Neder, 1964). Some representative tasks are:

—Removing propellant lines, transducers, test wiring, etc.
—Removing pressure-vessel bolts.
—Unclamping control-rod actuators.
—Unclamping and removing thrust structures.
—Removing bolts from turbopump flange and removing turbopump.
—And so on, until the fuel elements can be removed for detailed examination.

Engine-handling philosophy in the nuclear rocket program evolved directly from the Aircraft Nuclear Propulsion (ANP) Program, which also was concerned with large, hot engines (Layman, 1966). In addition, the disassembly tasks closely resemble those in the AEC's SNAP (Systems for Nuclear Auxiliary Power) reactor program. The major difference is size—a SNAP reactor has the dimensions of a waste basket rather than an automobile (Henoch, 1964).

In the acre-sized hot cells or "bays" used for aerospace reactor programs, small mobile manipulators can do many odd jobs, such as retrieving dropped parts unreachable by the main manipulators. Although the operating volumes of the large manipulators intentionally overlap, there is always the possibility that one will break down leaving parts of the hot cell inaccessible for a period. Mobile manipulators then come into action. Because of their usefulness, most large nuclear installations have one or more mobile manipulators (Drexler, 1965). The PaR–1 vehicle (Fig. 2.5) built by Programmed and Remote Systems Corp. is typical.

Between flights, a nuclear aircraft engine*—unapproachable because of induced radioactivity—would have to be serviced like other aircraft engines. A special vehicle, the Beetle, was developed during the ANP program for this purpose. Protected within a shielded cab, the operator could approach the engine and make limited repairs and adjustments. Because of its mobility and general purpose manipulator, the Beetle could also have been employed in crashes and other emergency operations.

---

* Although considerable development work was directed toward the construction of a nuclear aircraft engine (ANP Program), no operational engines were built.

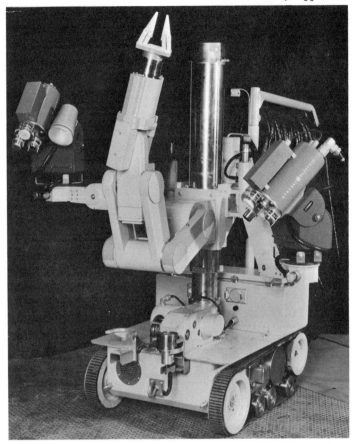

**Figure 2.5** The PaR-1 mobile manipulator. The vehicle is powered and controlled by cable. The manipulator arm and the two TV cameras are mounted on articulated booms. Height of the central support tube is 68 inches. (Courtesy of Programmed and Remote Systems Corp.)

### Nuclear Emergencies

Teleoperators are valuable in emergencies because they are mobile, versatile, dexterous, and relatively immune to environmental forces fatal to man. The same qualities that make them useful in rescue and salvage operations in space and under the sea carry over to the nuclear industry.

In a nuclear emergency, a teleoperator could enter the hostile environment, ensure that no further nuclear excursions could occur, measure radiation levels, reconnoiter the area, clean away debris (often with cable cutters, torches, etc.), and retrieve personnel (Briscoe, 1965).

Time is critical in a nuclear accident because radiation levels may kill survivors before a facility has been brought under control by shutting off electricity and fluids and fighting fires.*

Since nuclear incidents are rare, cleanup may also entail finding out exactly what went wrong. Debris must be recovered and much of it taken to hot cells for careful inspection. Finally, the facility must be decontaminated, a procedure involving sweeping, vacuuming, and washing with special chemicals. Time, of course, permits radioactive cooling, and removal of hot fuel and irradiated components further reduces radiation levels around the site of an accident.

Although a wide range of mobile manipulator systems exists (Drexler, 1965), none has been developed especially to deal with a major accident. Small mobile units like the PaR-1 can be helpful, but they are not designed for rapid entry via stairs, narrow passageways, and debris-cluttered floors. In a sense, using them is like using an ordinary automobile instead of a fire truck for fighting fires. As commercial nuclear power plants proliferate, specialized rescue vehicles—comparable in purpose to the Navy's *DSRV–1*—may be constructed.

So far, very few nuclear emergencies have occurred and development reactors have been intentionally located far from cities. In the now-cancelled ANP program, though, the AEC and Air Force pondered the possibility that a plane with hot engines might come down in a populated area, and three vehicles with rather strange names were built: the Bat, the Masher, and the MRMU (Mobile Remote Manipulating Unit). The Bat and Masher had no manipulators. The Bat was a shielded vehicle intended primarily for tractor operations, while the Masher boasted a crane. MRMU was a radio-controlled vehicle carrying two manipulator arms built by the Air Force Weapons Laboratory specifically for nuclear recovery operations. These vehicles are rarely used now.

The Alternating Gradient Synchrotron (AGS), at Brookhaven National Laboratory, on Long Island, has enough beam power to induce dangerous levels of radioactivity in the accelerator tunnel. Radiation levels occasionally exceed 100 roentgens per hour, precluding direct handling of the equipment. New accelerators now on the drawing boards will induce even higher radiation levels. Although most induced radioactivity decays rapidly with time, the time of a huge accelerator like the AGS is so expensive that downtime must be minimized. Consequently, Brookhaven has conceived of a master-slave manipulator that can quickly enter accelerator areas to repair and replace components or modify ex-

---

* One school of thought contends that personnel rescue must be consummated so rapidly that there would be little time to bring up teleoperator support. In this view, humans must enter the accident area and rapidly retrieve the survivors.

periments. A servo system for this manipulator has already been developed. Because many accelerator parts are fragile, Brookhaven adopted a force-reflecting servo manipulator scheme similar to that pioneered at Argonne National Laboratory (Flatau, 1969).

Some radiation-processing facilities also are expensive to operate. Although food, wood, plastics, and other materials usually go through irradiating zones on conveyor belts or automatic transport equipment, a need may arise for maintenance, repair, and modification of a production line without shutting down the source of radiation (reactor or radioisotope source). Teleoperators may turn out to be economically desirable in such facilities.

## TERRESTRIAL TRANSPORTATION

Once a vehicle leaves the smooth, hard, expensively prepared roadbeds that criss cross well-developed countries, wheels may become a liability, and legs may serve us better again.

Most walking machines built to date have been for development and demonstration purposes, although R. A. Liston reports that some crude draglines have been constructed employing walking machines (Liston, 1964). There is also a rather slow and ponderous walking machine in a German mine. These primitive machines, however, have been preprogrammed and therefore are not true teleoperators.

General Electric Company has carried out considerable study and development work on multilegged vehicles. Originally termed CAM's (Cybernetic Anthropomorphic Machines) or "pedipulators," such machines may replace men and animals on warfronts where roads (especially unbombed and unmined ones) are rarities. Walking machines also might be advantageous in swamp and polar exploration. Under the sponsorship of the U.S. Army and the Advanced Research Projects Agency, GE has been developing a prototype Walking Truck (Fig. 2.6). Such a vehicle might be applied in the future to off-road locomotion in military operations.

Are there also other ways in which teleoperators can aid soldiers? The so-called "man amplifier,"* an exoskeletal machine, can conceivably transform an ordinary soldier into a "super-soldier." A controllable, powered framework surrounding a soldier might amplify his strength and, at the same time, carry a protective shell. In effect, the soldier might become a walking tank, carrying a variety of heavy armament and still

---

* The term "man amplifier" was coined by Cornell Aeronautical Laboratory. A similar word, "maximan" has been coined by E. G. Johnsen to describe the teleoperator augmentation of man.

30  Teleoperator Applications

**Figure 2.6** Photograph of the Walking Truck prototype developed by General Electric for DOD. (Courtesy of General Electric Co.)

possessing much of the versatility and mobility of an individual. Cornell Aeronautical Laboratory, which pioneered exoskeletal work for the Navy, calls this the "servo soldier" concept. The exoskeleton can be magnified into an armored biped controlled by a man inside wearing a harness that communicates his arm and leg motions to the teleoperator. When man does not "wear" the machine, the machine is no longer a true exoskeleton, but rather a man-controlled walking machine and also a true teleoperator.

Exoskeleton work continues under the joint Army-Navy Project MAIS (Mechanical Aids for the Individual Soldier). One specific concept is "Hardiman," an exoskeleton enabling a man to lift up 1,500 pounds six feet in five or six seconds. Such a feat should even impress the Martians who invaded the Earth in walking machines in H. G. Wells' "War of the Worlds."

## ARTIFICIAL LIMBS

A good artificial limb is dexterous, general purpose, and operated by a man, and these are the key ingredients of this book's definition of a teleoperator.

A *prosthetic device* attempts to duplicate the functions of some missing part of the body. The tasks of a prosthesis are thus often those of a human hand, an arm, or a leg.

An *orthotic device* helps some weakened or atrophied part of the body to gain strength and dexterity. It does not replace a limb; a good example would be a powered exoskeletal brace to strengthen and steady a weakened arm. Training and exercising various parts of the body also are important applications of teleoperators. Medical engineering and teleoperator engineering overlap here.

Where engineering disciplines meet, intellectual cross fertilization often occurs. Medical engineering, for example, has developed ingenious joints, clever linkages, and sophisticated mechanical hands that grip harder when objects tend to slip. Aerospace engineering can provide better power sources, servomechanisms, and extensive knowledge of feedback control. Just where a more general and intimate confrontation will lead no one knows.

Teleoperators divorced from the body save for controls and sensors can also help people whose strength, freedom of motion, and dexterity are somehow limited. Feeding machines can be built wherein a specialized mechanical hand manipulates table utensils under direct control of the person being fed. Teleoperators could turn book pages, play cards, write, tune TV sets, and give the bedridden greater independence.

Teleoperators may eventually come to the surgeon's aid. At least three applications are now envisioned as teleoperators become more sophisticated:

>—Superclean surgery. Already operations have been performed with the patient completely enveloped by a sterile barrier of thin plastic. The plastic is so thin that the surgeon can work through it in glove-box fashion. Teleoperators of high dexterity and with considerably better touch feedback than now available could make truly aseptic surgery possible.
>—If clean surgery is feasible via a teleoperator, it is conceivable that a surgeon can operate from almost any distance. This idea is not

an unreasonable extrapolation of electrical master-slave manipulators with force reflection.

—Microsurgery is another target for teleoperators. Electrical circuits and mechanical devices can steady and scale down a surgeon's motions to any desired degree. Hand tremors can be damped out. With image magnifiers and intensifiers, work of great precision can be carried out in a way not too different from methods of connecting microelectronics circuits.

A more controversial application of teleoperators in medicine would be their use in the manipulation of the limbs and heads of brain-damaged children in a technique called "cross patterning." Experiments at The Institutes for the Achievement of Human Potential, in Philadelphia, have shown some improvements in the capabilities of such children through lengthy therapy of this type. Possibly teleoperators can supplant some of the lay therapists now employed; however, present thinking tends toward preprogrammed, computer-controlled machines rather than teleoperators.

One certain byproduct of the development of teleoperators and man-machine systems is a better understanding of the human body and its many subtleties. For example, the study of electromyography* for teleoperator control will undoubtedly lead to greater insight into the body's own control mechanisms.

## INDUSTRIAL APPLICATIONS

Accidental detonations sometimes occur when the constituents of explosives and rocket propellants are mixed, particularly during the development of new and unpredictable compounds. For many years technicians handled these powerful chemicals behind barricades with crude tongs and specialized mechanical devices (Rohm and Haas Co., 1961). As in nuclear work, the dexterity of these simple devices left something to be desired. Today, dozens of mechanical master-slave and unilateral manipulators, identical to those employed in hot cells, manipulate and blend hazardous chemicals.

Many chemicals are toxic or irritating when handled. These, too, could be handled by manipulators.

Normally, a glove box is adequate protection for biologists, although the least pin prick in the glove can be fatal in some work. In 1960, the Army's Biological Warfare Laboratories, Fort Detrick, Md., evaluated

---

* Electromyography is the study and utilization of the electrical potentials generated by muscle activity.

master-slaves and other remote-control devices to prevent personnel infection from biological agents and laboratory animals (Rawson, 1960). The Army ultimately concluded that more careful control of conventional techniques would be adequate. Nevertheless, the report remains an important part of the teleoperator literature because it is full of special and ingenious designs of teleoperator hands for handling animals, syringes, and other biological apparatus. Under severe circumstances, such as the environment following a biological attack, teleoperators might prove invaluable in decontamination and cleanup.

## Metal-Industry Potentialities

Teleoperators usually appear wherever the environment endangers man or the objects to be manipulated are too large or heavy for him. In forging operations, metal ingots are so hot that men cannot work close to them, and, even if they could, ingots are too heavy to handle manually. The obvious solution is a heavy-duty manipulator that can pick up a hot ingot, carry it to the forge, and manipulate it as desired (Hadfield, 1965). Some forging manipulators are permanent fixtures, but others are mobile. Capacities range as high as ten tons. Forging manipulators have little dexterity and are special purpose machines; therefore, it is stretching a point to call them teleoperators.

Another application where high temperatures favor teleoperators is the maintenance of high-temperature furnaces. Here, heat-resistant teleoperators could enter the furnaces long before men could, inspect the interior, and make repairs where necessary. Furnace downtime would be minimized.

High-vacuum production processes might benefit if man's dexterity could be transferred through vacuum chamber walls. High-vacuum welding and high-vacuum metal production both require deft operations that man could do with the help of teleoperators. Downtime for maintenance and repair could be reduced. There is a close economic parallel between this application and the use of manipulators in high-altitude test chambers in the aerospace industry.

It is intriguing to apply the teleoperator concept to fabrication and maintenance problems in industry. A highly flexible arm can explore and manipulate in spaces so tortuous and confined that human arms are completely barred. Aircraft welding, the cleaning of pipes and retorts, and searching for broken bits in drill holes are but a few possibilities.

## The Electronics Industry

In the old days, a radio amateur could build a passable rig in his basement—even in poor light and next to the coal bin. Now rows of women in dust-free garments assemble electronic parts under microscopes in clean rooms. Cleanliness and miniaturization beckon teleoperators. The electronics clean room workers combat dirt and airborne contaminants because solid-state components are notoriously sensitive to impurities. Welds and solder joints, too, suffer in the presence of dirt. In fact, the lure of higher performance may eventually place most microelectronics and integrated circuit construction in a vacuum or controlled atmosphere. Micromanipulators worked by personnel outside the "superclean" room may then assemble and fabricate the desired equipment. Most of the micromanipulators employed by the electronics industry today, however, are special purpose tools with little dexterity. They operate from controls like those on lathes and other machine tools and have few of the attributes of the human hands. The large numbers of repetitive operations make tool specialization profitable here.

### Construction and Mining

Steeplejacks, sand hogs, and skyscraper riveters have romantic but hazardous jobs that teleoperators could do. Men still do such work because teleoperators are expensive to develop.

Mining has become less dangerous in recent years; excavating machines have sent much of the work force back to the surface. Tunneling is still hazardous and time consuming. When explosive charges are placed, men and machines retreat before the detonation and move back in gingerly afterward. Conceivably, a heavily armored teleoperator could be constructed that would continuously place drill charges and detonate them against the working surface. It would then leave behind a path of suitably fragmented material for supporting mucking machinery to convey back to the minehead. Furthermore, there would be no need for ventilation and other provisions to support and protect fragile men.

## PUBLIC SERVICE APPLICATIONS

Armored, superstrong policemen and firemen have been suggested (mostly in jest) by more than one engineer. Super-criminals may well appear first; they have on television. A fire, whether in a warehouse or forest, poses an environmental threat that a teleoperator can counter with its great resistance to heat and independence of a breathing atmosphere. No teleoperator has yet been designed for this purpose, but tasks, such as hose handling, application of chemicals, preparation of firebreaks, and so on, are easy to imagine.

Public service officials also must deal with spills and releases of toxic gases and fluids. Releases of chlorine, for example, have frightened many communities. Truck and train wrecks have often spilled noxious substances in populated areas. Perhaps someday a general purpose teleoperator will be built to cope with such situations without endangering man.

In February 1966, the Chicago papers related how a lipstick-sized capsule of radioactive cobalt was accidentally dropped at the Lutheran General Hospital. The technicians loading the source into its container fled and received only a small radiation dose. To retrieve the cobalt source, personnel from Argonne National Laboratory ran a PaR-1 mobile manipulator into the area, picked up the source with the manipulator, and dropped it in its lead container. A teleoperator was the hero in this mishap.

Safeguards stipulated by the AEC have prevented undue exposure of the public to radiation. As we progress farther into the atomic age, however, a state or large city may find it worthwhile to add teleoperators to its line of emergency vehicles to deal with nuclear accidents.

Perhaps the greatest emergency is the cleanup of the mess man has made of his environment, particularly his cities. In fact, man has ahead of him the greatest construction job he has ever faced. Already some futurists are talking about scrapping whole cities and rebuilding them with man in mind and not industry. The factories can be consigned underground where robots and teleoperators will do the dirty work. Above ground, man freed from mindless, demeaning jobs can build—again with the help of teleoperators—any sort of utopia he wishes, unconstrained by seas of grinding, smoking machines. With the sight and sound of TV augmented by teleoperators' touch and dexterity, man can build anything he wishes anywhere he wishes—without getting caught in the rush-hour traffic.

## FROM PUPPETS TO SERVANTS

Aboard the Santa Fe and Disneyland Railroad, passengers can see lifesize ostrich dinosaurs drinking from a vanishing waterhole. Other prehistoric monsters search for food and fight among themselves. The monsters are preprogrammed and controlled by Disney's Audio-Animatronics system. By removing the preprogramming limitation, some P.T. Barnum of the future can fill parades and circus rings with giants, monsters, and gladiators that duel to the death. Indeed, combat by teleoperator might become a fad like the current "crash" contests between jalopies. Instead of manipulating arms and legs by strings like the puppets of yore, electromagnetic and audio signals bring life to these machines.

What a status symbol a walking-machine golf caddy could be! To future generations no safari or mountain-climbing expedition may seem complete without teleoperators to clear the trail, carry supplies, and tote the elephant tusks back to camp.

The "Far, Far Out" category of concepts includes the Man Multiplier or "Doppelgang," in which one man controls tens or even thousands of identical machines, all making the same motions simultaneously in concert with the human operator. The reader's imagination may generate applications for this idea.

The "Miniature Man" concept is also remote, although there is no fundamental reason why teleoperators cannot be built much smaller than man as well as larger. Several imaginative scientists have toyed with the idea of building a teleoperator able in turn to build a smaller replica of itself, and so on, smaller and smaller, until the descendants reach atomic dimensions. Science fiction, yes; but all vital fields have their wild frontiers, and teleoperators are no exception.

# III
# SUBSYSTEMS AND THEIR INTEGRATION

The arms, legs, and hands of a teleoperator inevitably attract the most attention because they are the most nearly human portions of the machine. Yet, to fulfill man's objective in outer space, under the sea, and elsewhere, a teleoperator must be capable of propelling itself from place to place, communicating its position and operational status to man, and, most important, effectively projecting man's presence into the environment being explored. The complete teleoperator, therefore, has an array of subsystems that make it a sentient, mobile, and hopefully, profitable extension of man.

When teleoperator complexity greatly exceeds that of primitive unilateral manipulators, conceptual visualization becomes easier if the system is broken down into subsystems. Any such dissection is arbitrary, but the subsystems listed below have proven useful in teleoperator analysis and design. The ten teleoperator subsystems can be defined in terms of functions and typical hardware:

—*The actuator subsystem* carries out manipulations, walking, and other dexterous activities ordered by the human operator. The actuator subsystem is the "effector" portion of the teleoperator system. The slave arms and hands of the familiar master-slave manipulators are typical actuator subsystems.* Yet, it is too restrictive to imagine actuator subsystems as always anthropomorphic. (See Table 3.1 for definitions.) Wrist extension, unlimited wrist rotation, and lack of elbow points already make some master-slave manipulators nonanthropomorphic to a degree. Tomorrow may see suction grips, telescoping legs, and many-jointed arms; viz., the Serpentuator concept. Of course, teleoperator actuator

---
* The motors and other devices that *create* motion are often called "actuators."

## Subsystems and Their Integration

Table 3.1 Definitions of Some Common Types of Teleoperators

| Type | Definition used in this book |
|---|---|
| Unilateral teleoperator | A teleoperator in which force and motion can be transmitted only from the operator controls to the actuators. |
| Bilateral teleoperator | A teleoperator in which force and motion can be transmitted from the operator controls to the actuators and vice versa; i.e., the slave arm can move the master arm. (Note: "bilateral" does not imply physical symmetry here as it does in biology.) |
| Rectilinear teleoperator | A teleoperator possessing several degrees of freedom in rectangular coordinates. Generally, these degrees of freedom are associated with overhead bridge-crane positioning systems. "Rectilinear" is often used incorrectly as a synonym for "unilateral." Joints with angular freedom are often termed "polar" in the literature. |
| Master-slave teleoperator | A teleoperator in which forces and torques are proportionally reproduced from the controls (master) to the actuators (slave). A master-slave is bilateral in at least seven degrees of freedom in each arm/hand. All degrees of freedom can be controlled naturally and simultaneously. This term was originated at Argonne National Laboratory. |
| Anthropomorphic teleoperator | A teleoperator with controls and an actuator subsystem resembling the human body. An exoskeleton *must* be anthropomorphic to a large extent; many manipulators possess fingers, wrists, and shoulder joints, etc. At best, this is a vague and relative term. |

subsystems may also be stronger and more precise than man's limbs and hands.

—*The sensor subsystem* is the sentient portion of the teleoperator. It may see, feel, hear, smell, or otherwise sense the environment, giving the operator rapport with transactions in the actuator space. More than any other subsystem, the sensor subsystem enables man to project himself across distance and through barriers into the working area. Television cameras, microphones, piezoelectric pressure pickups, infrared cells, sonars, and navigation gyros are only a few of the possibilities. Like the actuator subsystems, many sensors are nonanthropomorphic. The sensor subsystem also tells the operator the "status" of the teleoperator by relaying data on vehicle location, velocity, attitude, and the system operational mode.

—*The control subsystem*, including, of course, the human operator, analyzes the information fed back by the sensor subsystem and prepares new commands to the various subsystems. In the most obvious case, the operator sees an object and moves controls that cause a manipulator to pick it up or otherwise manipulate it. Or, a status indicator may signal that an attitude-control actuator is not functioning, causing the operator to take corrective action. The purview of the control subsystem extends far beyond the master portion of a master-slave manipulator or the switch-type controls of a unilateral manipulator. To illustrate: since a teleoperator is mobile in the generalized case, the control subsystem also receives and analyzes navigational information and dispatches appropriate commands to the propulsion and attitude-control subsystems. The control subsystem is the teleoperator's brain, decision maker, and command generator.

—*The communication subsystem* is the nervous system of the teleoperator. To it and from it speed all data and commands. Hard wire, electromagnetic, sonar, and mechanical links tie all of the subsystems to the control subsystem in those cases where the operating space cannot be seen directly by the human operator. When hot-cell windows and submersible portholes permit direct visual access, the data-handling capacity of the communication subsystem is augmented by a visual channel of great bandwidth. Direct vision represents a superlative communication link.

—*The computer subsystem* aids man in controlling the teleoperator. In this function, the computer converts incoming information into displays that the operator can easily comprehend. It makes calculations and predictions to support and improve decision-making by the operator. Further, the computer may relieve the operator's burden by storing command subroutines that can handle the more perfunctory teleoperator tasks. For example, stowing the manipulator on a submersible can be carried out entirely by stored subroutines. In distant (viz., planetary) operations, where signal time delay and bandwidth are restrictive, a small computer in the actuator space can compress data for transmission back to the operator. Hopefully, this same computer can also give the teleoperator some degree of autonomy and quick reaction. (See later discussion in this chapter.)

—*The propulsion subsystem* may comprise rockets, motordriven wheels, screws, or leg-like parts of an exoskeleton, depending upon the application.

—*The power subsystem* provides electrical, hydraulic, mechanical, and other forms of power to the various subsystems. The energy source may be man himself (as in mechanical master-slaves and some prostheses), a battery, a solar-cell bank, compressed gas, an internal combustion engine, etc.

—*The attitude-control subsystem* employs jets, propellers, electromagnets (in space), telescoping structures, docking arms, and a variety of other devices to stabilize and control the spatial orientation of the teleoperator. In some cases, the actuator subsystem itself may provide the necessary forces for attitude control. Commands for attitude control may come directly from the operator, but very often the operator will be short-circuited by local control loops, such as those used for maintaining a satellite's Earth orientation.

—*The environment-control subsystem* maintains temperatures, pressures, atmospheric composition, and other environmental parameters within specified limits. Heaters, cooling elements, and various kinds of life-support equipment are available for these functions. Like attitude control, a suitable environment is usually maintained, without conscious effort on the part of the operator, through the use of local control loops; viz., thermostat-controlled electronic compartments.

—*The structural subsystem* unites and supports other subsystems. In teleoperators, of course, the system is, as a whole, often divided physically by an environmental barrier or by great distances. In orbital and deep-sea missions, the operator commonly resides within the teleoperator vehicle, but this is certainly not a necessary arrangement.

Schematic isolation of each subsystem from the teleoperator-as-a-whole aids the engineer by setting before him limited sets of related functions. It is easier to grasp and visualize the hardware form of the communications subsystem, for example, when it lies separated from the complexities of the overall system. Balancing this advantage is the problem of glueing the separated subsystems back into a viable, unified system.

## SUBSYSTEM INTERFACES

In practice, no one ever designs a subsystem without thought for the overall system and the objectives that have been assigned to the teleoperator. Design cannot proceed on the basis of admonishments alone; teleoperators are too complex for that. The so-called systems approach" disciplines the conceptual designer and the applications engineer alike. The systems approach permits the luxury of subsystems excision without overall system degradation caused by poor interface matching when the subsystems are reassembled.

The first step in the systems approach is the definition of the system and its component subsystems—something just done for teleoperators. Next, the performance of the teleoperator must be expressed in terms of some overall figure of merit. In military systems, the over-riding figure of merit is often "cost effectiveness." For a manipulator engaged in some sort of production activity, the figure of merit might be measured in fuel elements handled per hour, or perhaps in more abstract terms as the time taken to assemble a standardized test object by a skilled operator. The speed and versatility of various manipulators can be compared on a standard basis if such a scale of value can be established. Of course, cost, maintenance requirements, and reliability are also important. The point here is that objective design of any complex system requires some definition of excellence that can be optimized by varying system parameters and, in turn, subsystem design. Needless to say, much design work, some of it excellent, proceeds with a lot less objectivity than that afforded by systems analysis.

Assuming the value of the systems approach in teleoperator design, we are immediately faced with the unsettling fact that nearly all teleoperators are applied in non-routine, non-standard operations that are not easily characterized by some single figure of merit. What is the figure of merit for a teleoperator prowling the Martian surface or a deep-sea rescue vehicle retrieving crewmen from a sunken submarine? Teleoperators are valuable because, with man in the loop, they can cope with unpredictable, unmeasurable events. The versatility that makes teleoperators valuable also makes them difficult to analyze.

If an overall figure of merit can be conceived and formulated in terms of subsystem parameters, the establishment of subsystems specifications is easy. The subsystem is designed to the range of parameters that optimizes the performance of the whole teleoperator system. Without guidance from systems analysis, engineers resort to intuition and experience. Most teleoperators move from concept to operational status via this road precisely because they are generalized machines rather than specialized systems that can be optimized to do a specific job.

Experience and intuition, if they are to guide subsystem integration, must be formulated verbally and shared among engineers. Let us take a specific example to see how this can be done. A power subsystem designer may be asked to provide a package that will yield a kilowatt of electrical power for six months and weigh less than 1,000 pounds. Environmental conditions and other parameters must also be specified if the power plant is to work properly when the subsystems are all assembled. Superimposed on these *subsystem specifications*, drawn most likely from rough feasibility studies, certain guidelines or *design philosophies* are also set down. One well-established design philosophy in manipulator design is that of *spatial correspondence*; that is, a motion in the control space should be duplicated in the actuator space. This is a design consideration that depends to a large extent upon the type of work to be done. Nevertheless, it has considerable value to a manipulator designer over and above narrow specifications such as lifting capability, reach, and so on.

System performance specifications, when interpreted as subsystem specifications and design philosophies, figure critically in reuniting subsystems into an effective whole. Still another kind of specification tells us more about the inner workings of the integrated teleoperator. This is the *interface specification*. Very succinctly, the interface specification tells the designer just what interface conditions—voltages, heat fluxes, data rates, etc.—he will have to provide if his subsystem is to mesh neatly with the nine adjacent subsystems.

Nine important interface "forces" exist in any teleoperator. The thermal interface specification allows the designer to bridge the thermal interface between, say, his power plant and the communication subsystem; it may stipulate specific temperatures and heat fluxes on the exterior of the power plant in such a way that they will not compromise the sensitive electronic gear in the communication subsystem. Examples of other interface forces are given in Fig. 3.1.

Between the ten teleoperator subsystems are $10 \cdot 9/2 = 45$ interfaces, each bridged by a possible nine types of interface forces. Obviously, all interface forces are not important in teleoperator design. A little thought weeds out the trivialities. Of course, the importance of some of these interface bonds varies with application. To illustrate, the mechanical interface between the control and actuator subsystems is vital in mechanical master-slave manipulators in which the operator's motions are communicated directly to the slave by cables and metal tapes. This interface does not exist in the electrical master-slaves.

While the necessity of matching electrical interfaces, such as voltage and current, with the power subsystem are manifest to all engineers, the *lingua franca* of the more advanced teleoperators is the *data word*. Subsystems converse among themselves by means of data words; the operator commands the actuators to perform dexterous operations with data words; the whole commerce of information exchange moves via the data word. The information interface is critical to successful intrasystem communication. Bit rate and word format, i.e., length, bit arrangement, etc., are highly important interface parameters.

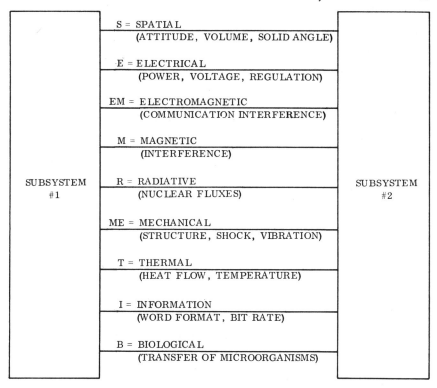

**Figure 3.1** Nine different kinds of interfaces exist between any two teleoperator subsystems. The actual importance of each kind of interface depends upon the subsystems involved.

Somewhere in the design process, interface specifications must be issued to all subsystem design groups; they are probably the most powerful tools promoting smooth hardware integration.

# IV
# TELEOPERATOR DESIGN PRINCIPLES

A design philosophy consists of general guidelines that summarize succinctly both hardware experience and theoretical expectations. It does not include specific performance goals, such as a particular lifting capability or level of power consumption. Rather, a design philosophy transcends specific missions, special applications, or a given type of teleoperator. A few important guidelines, however, will always be application-specific, such as the well-known admonition to use only radiation-resistant materials in hot-cell teleoperators. Finally, design philosphies are not hard-and-fast rules that have to be met with the rigor that engineers associate with design specifications; they are road signs, strategies, and distillations of experience. As such, they can be disregarded or modified at times, particularly when designing teleoperators for radically new environments or applications.

First, we will delineate those design philosophies that apply to all teleoperators and, after that, those few that are specific to the various application areas defined in Chapter 2. The more important general design philosophies fall rather neatly into three categories:

1. Those that ease the burden on the human operator.
2. Those that make the teleoperator a more effective machine.
3. Those that extend the teleoperator lifetime.

In the first category are such suggestions as:

1. The positions and velocities of teleoperator actuators should resemble those of the controls to help the operator project his presence into the actuator space. This is the well-known principle of "spatial correspondence." Most master-slave manipulators adhere closely to this philosophy, although controls on some space vehicles and submersibles may be scaled down in size to save volume. Even unilateral manipulator

designers adopt the philosophy of spatial correspondence when they coordinate their switch and joystick controls with the manipulator motions; i.e., pushing a switch left makes the manipulator arm swing left. Note that this principle applies to anthropomorphic and nonanthropomorphic teleoperators alike.

2. Teleoperator actuators should be modeled after man so that the operator will feel closely identified with the arms, hands, and fingers he is activating. (This question of anthropomorphic vs. nonanthropomorphic teleoperators will be discussed in Chapter 5.) At times, this philosophy can be waived to advantage. Most people, for example, feel at ease driving an automobile despite the nonanthropomorphic controls, actuators, and sensors. True, a car is not a teleoperator, but the illustration suggests that man is a more pliable component in the man-machine arrangement than is generally believed, and may not always have to be pampered.

3. Vision, force reflection (or "feel"), and all the environmental factors a human can sense should be incorporated into teleoperator design. The objective of "sensory correspondence" is also to enhance the operator's identification with the task at hand. Contemporary teleoperators rely primarily on vision, because the cost of adding sound and feel may not be commensurate with the improved effectiveness of the teleoperator. Sensory correspondence, like all other design factors, must be balanced against other desirables.

4. Teleoperator controls should not be "spongy" or sluggish, yet they should not be so sensitive that the operator's least tremor is communicated to the actuators. The automobile analogy is apt again—the steering wheel should have a little but not too much "play" in it. Drift must be negligible, too. Force feedback in the teleoperator should be clean and crisp but not so strong that it tires the operator. (Mechanical and electrical force multiplication can reduce force feedback to tolerable levels.)

5. The visual scene communicated to the operator should be immobilized; that is, spatially fixed. As the operator turns his head, he should see a different portion of the environment. We could call this "visual correspondence" and define it as a partial union of sensory and spatial correspondence. It means more than merely a faithful, picture-like reproduction of the scene in the operating space. Today, only head-controlled television sets and large hot cell windows can create visual correspondence.

6. Actuators and optical sensors should not mutually interfere, that is, the manipulator hands should not obscure the operator's view of the object that he is manipulating.

7. All actuator degrees of freedom (joints, wrist extensions, etc.) should be designed to move continuously and simultaneously, without excessive backlash, much like the operator moves his limbs and digits. It is tempting to add that there should be no cross coupling between different degrees of freedom; viz., movement of joint X does not cause some motion of joint Y as well. This, strangely enough, would be a nonanthropomorphic requirement because human tendons are often "cross-coupled." Zero cross coupling makes control theory simpler, but it is not always essential to good hardware design.

In the second category of design philosophies are those that make the teleoperator more useful or effective.

8. Teleoperator design should be kept "generalized" as far as possible. Biologists maintain that the human being is successful among the animals because his brain, limbs (excluding the feet), hands, and other "subsystems" can perform many different functions; i.e., they are unspecialized. Since teleoperators are extensions of man, they will be of greatest general value if specialization is avoided.

9. The actuators should exhibit "compliance" or compatibility in degrees of freedom with the motions making up the mission. If the job involves rotary motion, such as turning bolts, rectilinear manipulators are seldom desirable. Compliance means matching the teleoperator to the job. Compliance, rather obviously, implies specialization of teleoperator design, contradicting the preceding design suggestion. Such conflicts are inevitable in engineering any complex system. Trade-off studies must be made to determine what mix of compliance and generalization yields the highest performance over the expected application spectrum.

The third and final group of philosophies includes those that help the teleoperator survive the rigors of use and environment.

10. Teleoperator design should be clean and simple, with the most critical components paralleled to encourage high reliability. This sounds like an unnecessary hortatory remark, but reliability cannot be overemphasized in environments where recovery and repair are difficult or impossible.

11. Self-repair capability should be built into a teleoperator that cannot be repaired by man directly. Most teleoperators have arms and hands backed by human operators. With this dexterity and resourcefulness available, defunct parts can be replaced if spares and proper tools are within reach of the manipulator hands. The teleoperator should be designed with an eye to easy disassembly and repair by its own arms and hands. In effect, this means that the manipulator arms should be able to

reach all repairable components, and that the viewing system should be adjustable to make the teleoperator introspective. Two manipulator arms are particularly useful in self-repair situations.

12. Closely associated with self-repair is the "modular" concept, wherein the teleoperator is constructed from easily replaceable building blocks. Maintenance and self-repair are easier then and improved com-

Table 4.1 Application-Specific Design Philosophies

| Application area | Design philosophy |
|---|---|
| Aerospace | Teleoperator-bearing space vehicles should be attached and anchored firmly to the target to preclude excessive attitude perturbations. |
| | Local computers should be incorporated in teleoperators on lunar and deep-space missions to provide supervisory control (see Chapter 5). |
| | Preview control or its equivalent should be employed in planetary teleoperators to overcome time-delay problems (see Chapter 5). |
| | Low weight and power consumption, and high reliability are critical. |
| Undersea | Teleoperator attachment and anchoring are required, as described above. |
| | Materials should be compatible with seawater. |
| Nuclear | Teleoperator components in the actuator space must be radiation-resistant. |
| | Hydraulic manipulators should be avoided in hot cells because of the great difficulty in cleaning up oil leakage. |
| | Low cost is a critical factor in commercial application. |
| Terrestrial transportation and materiel handling | Low cost is important because of competition with helicopters and wheeled vehicles. |
| | Gaits that annoy or sicken the operator must be avoided. |
| Medical | Prostheses must "look right." |
| | Low cost is essential for prostheses. |
| | Equipment for surgery must be able to withstand sterilization. |
| | Low weight and power consumption are essential. |
| Chemistry and biology | Teleoperator actuators must be able to withstand repeated cleaning and, in some cases, sterilization. |
| | Precision motion is desirable. |
| Public service | Low cost is an important factor because of the competition of conventional equipment. |
| | Rugged construction is essential. |
| Entertainment | Operator concealment is often an important factor. |

ponents can be installed when developed; viz., more powerful or longer arms.

13. The teleoperator should be provided with a stable environment insofar as possible. Temperatures, the internal atmosphere, vibration loads, and so on, must be controlled carefully if long life is desired. In practice, this idea is translated into environment and interface specifications that are consistent with known lifetime characteristics of the teleoperator components. Unfortunately, little reliability data is available on teleoperator components.

14. The teleoperator actuator subsystem should be provided with proximity and limit switches as well as stress-limiting devices, such as slip clutches and pressure valves. With some foresight, the designer can prevent teleoperator damage that might otherwise be incurred in trying to lift or move overweight objects, or by collisions among its own parts and the targets being handled.

Some of the more important specialized guidelines are summarized by application area in Table 4.1.

Concluding this section is a second table in which ten teleoperator subsystems are cross-indexed with the eleven important application areas. Table 4.2 is a preview of the rest of this chapter as well as of Chapters 5, 6, and 7; it summarizes key subsystems.

The manipulators, which are often manlike, and the sensory organs that attempt to duplicate the scene a man would see if he could occupy the same space as the actuators, often are so critical to the success of a teleoperator that a full chapter has been assigned to each; these are Chapters 7 and 6, respectively. Another critical subsystem chapter is Chapter 5, which deals with the critical controls that mediate between man and machine. The remaining seven subsystems are discussed in the remainder of this chapter.

## THE COMMUNICATIONS SUBSYSTEM

The teleoperator communications subsystem carries information among all subsystems. The heaviest traffic is from the sensors to the control subsystem and from the control subsystem to the actuator subsystem. There often are, however, numerous communication channels that "short-circuit" the control subsystem completely, such as those that aid in automatic temperature stabilization and those that improve grip control in advanced artificial hands. These local "loops" are analogous to the systems of human nerve fibers that transmit the reflex signals that bypass the brain. Local signals are carried from point to point within the teleoperator by "hardwire"; that is, ordinary electrical wires and cables. Hardwire

Table 4.2 Comparison of Teleoperator Subsystem Features by Application Area

| Application area | Actuator subsystem |
| --- | --- |
| Aerospace | Electrical master-slaves will probably be best in space. Mechanical master-slaves now used in terrestrial test chambers. |
| Undersea | Electrohydraulic and electric unilateral manipulators now dominant. |
| Nuclear | Mechanical master-slaves abundant; electrical master-slaves at ANL. Electrical unilateral manipulators commonly used in very large hot cells and on vehicles. |
| Terrestrial transportation and materiel handling | Walking machines will usually have more than two legs because of the stability requirement. Unilateral actuators for military use. Man amplifiers will probably be hydraulic and electrohydraulic. |
| Medical | Wide array of limbs, hands, and ingenious joints and linkages now available. |
| Chemistry and biology | Mechanical master-slaves dominant. |
| Metal industry | Heavy-duty hydraulic unilateral manipulators used almost exclusively. |
| Electronics | Mechanical and electrical unilateral and master-slave manipulators will probably be employed. |
| Construction and mining | Heavy-duty hydraulic unilateral manipulators used extensively. |
| Public service | Unilateral manipulators probably will dominate this field. |
| Entertainment | Both mechanical master-slaves and electrical unilateral devices will probably be used. |

Table 4.2 Comparison of Teleoperator Subsystem Features by Application Area (Continued)

| Application area | Sensor subsystem | Control subsystem |
|---|---|---|
| Aerospace | Direct viewing can be used in orbital work; TV for lunar and planetary work; force reflection likely in both applications. Direct vision in terrestrial test chambers. | Closed-loop tracking by operator likely in orbit. Preview display and supervisory control for distant planets. Open-loop control reasonable out to Moon. |
| Undersea | Direct viewing from submersibles. TV for unmanned exploratory craft and rescue vehicles. Sonic imagers may find use where vision is difficult. | Open and closed-loop operator tracking used. Miniaturized electrohydraulic position controllers becoming common. Switches and joysticks for the now-dominant unilateral manipulators. |
| Nuclear | Direct vision and force feedback dominant in hot-cell work. TV employed on mobile equipment and in large hot cells. Microphone pickups common. | Open and closed-loop operator tracking. Switches, joysticks, master-slaves, exoskeletal control devices. |
| Terrestrial transportation and materiel handling | Direct vision. | Closed-loop, operator-tracking. Exoskeletal controls. Subroutines for easy terrain. |
| Medical | Direct vision for prosthetics. TV for remote surgery inescapable. | Closed-loop, operator tracking. Various body-operated switches and exoskeletal controls. Myoelectric control under development. |
| Chemistry and biology | Direct vision. | Closed-loop, operator tracking. Master-slave controls. |
| Metal industry | Direct vision. | Closed-loop, operator tracking. Switch controls. |
| Electronics | Direct vision. | Closed-loop, operator tracking supplemented by subroutines. Switch controls. |
| Construction and mining | TV. | Closed-loop, operator tracking. Switch controls. |
| Public service | Direct vision supplemented by TV for mobile equipment likely. | Closed-loop, operator tracking. Switch controls. |
| Entertainment | Direct vision. | Closed-loop, operator tracking, supplemented by subroutines. Exoskeletal controls. |

Table 4.2 Comparison of Teleoperator Subsystem Features by Application Area (Continued)

| Application area | Communication subsystem | Computer subsystem |
|---|---|---|
| Aerospace | Electromagnetic links for distant teleoperators inescapable. Hardwire links for orbital manned work capsules. Mechanical manipulators for test-chamber work. | Digital computers may be employed in local preview control, in distant supervisory control, and in data compression. |
| Undersea | Hardwire links, including trailing vehicular cables, are most common. Acoustic links possible. | Digital computers may be used for distant supervisory control, and in data compression. |
| Nuclear | Mechanical links dominate in master-slave type of manipulators. Cable and radio-controlled vehicles exist. | None. |
| Terrestrial transportation and materiel handling | Hardwire links for unilateral manipulators. | None. |
| Medical | Mechanical and hardwire links in prostheses. | None. |
| Chemistry and biology | Mechanical links. | None. |
| Metal industry | Hydraulic links most common. | None. |
| Electronics | Hydraulic and hardwire links. | None. |
| Construction and mining | Hydraulic links for manned machines. Radio and hardwire links for vehicles. | None. |
| Public service | Hardwire links. | None. |
| Entertainment | Acoustic, mechanical, radio, and hardwire links. | None. |

Table 4.2  Comparison of Teleoperator Subsystem Features by Application Area (Continued)

| Application area | Propulsion subsystem | Power subsystem |
|---|---|---|
| Aerospace | Reaction engines (chemical or cold gas) for space. Walkers and wheels for planetary surfaces. | Chemical APU's, fuel cells, solar cells; nuclear power in the future. |
| Undersea | Screws and jets now used in submersibles. Tracks for bottom crawlers. | Batteries, chemical APU's nuclear power plants, and electric lines now in use. |
| Nuclear | Bridge-crane-type carriages. Tracks used for most vehicular manipulators; wheels on a few. Walkers possible in future. | Human-powered masterslaves, electric lines, chemical engines (gasoline, Diesel), all in use. |
| Terrestrial transportation and material handling | Walking machines likely. | Chemical engines (gasoline, Diesel, gas turbines) for future walking machines and exoskeletons. |
| Medical | Only walking machines and artificial legs considered. | Human power, compressed gas, batteries now in use. |
| Chemistry and biology | None. | Human-powered masterslaves used. |
| Metal industry | Heavy tracked or wheeled vehicles. | Chemical engines (gasoline, Diesel). |
| Electronics | None. | Human-powered masterslaves, electrical lines. |
| Construction and mining | Heavy tracked and wheeled vehicles. | Chemical (gasoline, Diesel). |
| Public service | Heavy tracked and wheeled vehicles. | Chemical (gasoline, Diesel). |
| Entertainment | Walking machines. Some wheeled vehicles. | Human power, electric lines. |

Table 4.2 Comparison of Teleoperator Subsystem Features by Application Area (Continued)

| Application area | Vehicle attitude-control subsystem | Environmental-control subsystem | Structure subsystem |
|---|---|---|---|
| Aerospace | Cold and hot-gas jets, docking arms, and gyros will probably be employed. | Active (moving) and passive radiators; subliming and evaporating materials; various heat sinks and various life-support systems have all been proposed. Meteoroid and radiation shields. | Space capsules proposed for manned orbital systems. Open frames and polygonal shells suggested for unmanned vehicles. |
| Undersea | Screws now. Motion of operator and/or fluids inside submersible used to some extent. Docking arms potentially useful. | Seawater heat sinks. Various life-support systems. | Massive hulls to withstand extreme pressures. Open frames for unmanned vehicles. |
| Nuclear | None. | Vehicle radiators. Radiation shielding. | Master-slaves suspended from central supports. Column-and wall-mounted unilateral manipulators. |
| Terrestrial transportation and materiel handling | Walking-machine legs stabilize operator. | Vehicle radiators. Armor in warfare. | Exoskeletons, legged platforms proposed. |
| Medical | None. | None. | Artificial limbs with internal or external skeletons. |
| Chemistry and biology | None. | None. | Master-slaves suspended from horizontal support. |
| Metal industry | None. | Vehicle radiators. | Truck/tank structures. |

Table 4.2  Comparison of Teleoperator Subsystem Features by Application Area (Concluded)

| Application area | Vehicle attitude-control subsystem | Environmental-control subsystem | Structure subsystem |
|---|---|---|---|
| Electronics | None. | None. | Master-slaves suspended from horizontal support. |
| Construction and mining | None. | Vehicle radiators. | Truck/tank structures. |
| Public service | None. | Vehicle radiators. | Truck/tank structures. Legged platforms. |
| Entertainment | None. | None. | "Internal" skeletons. |

communication is, in fact, the only type of data link employed to any significant extent in today's teleoperators in addition to the inescapable mechanical tapes, cables, and gears.

What kinds of communication links are physically possible? We consider here only the channels between the control subsystem and the other subsystems, not only because they dominate the communication picture, but also because it is here that the greatest variety of links exists. This fecundity doubtless results from the frequent separation of the human operator by large distances and/or physical barriers.

If the human operator is physically close to the actuator, mechanical and hydraulic communication links are simple and reliable. When separation distances exceed a dozen feet or so, electrical cables (hardwire) supplant mechanical and hydraulic links. Cables are technically feasible up to distances of perhaps a few miles, particularly in undersea applications where radio communication is difficult or unfeasible. Beyond the practical range of cables, the intervening medium must carry the signals. In outer space, there is little choice except trains of electromagnetic waves; i.e., radio or light signals. Beneath the sea, acoustic communication links are possible, although relatively undeveloped. The choice of the communication system thus depends upon distance and the regime of application. Current solutions are summarized in Table 4.2. (One important communication link not mentioned in the preceding discussion is provided by reflected light waves "modulated" by the scene in the operating space. Since direct viewing is intimately connected with the subject of sensor subsystems, its full discussion has been deferred to Chapter 6.)

The basic commodity of communication is information. We want to transmit it without distortion, without the addition of noise, and as cheaply as possible (Krassner, 1964). Distortion and noise cannot be completely eliminated, however, because the medium itself and the communication equipment introduce perturbations beyond the control of the designer. Information is a commodity that may be treated mathematically in a way similar to the state variables employed in thermodynamics. No matter how hard the engineer tries, perfect transmission of information, like a 100 percent efficient heat engine, is impossible. Not surprisingly, the more nearly perfect communication is made, say, through the use of redundant and error-correcting codes, the more expensive each piece of information (the bit) becomes. "Expense" in a communication system is generally measured in terms of bandwidth or power required.

In teleoperator design, the problems of noise, bandwidth and power are particularly acute. On the "command" portion of the link, dozens, perhaps scores of degrees of freedom must be controlled smoothly and with precision. This implies a very wide bandwidth. A whole experiment may be jeopardized if noise or a "bit error" is somehow introduced into the link. On the return or data portion of the link, environment sensor information is likely to be video (TV), which also demands a wide bandwidth. The so-called "status" information that tells the operator the positions, velocities, and applied forces for each degree of freedom and the "health" of the teleoperator needs considerable bandwidth. Transmitter power can solve bandwidth and noise problems in a brute-force sort of way, but hostile environments generally make power a scarce commodity.

Teleoperator commands and the returning sensor signals may be analog or digital. The teleoperator's present state of development makes inter-subsystem communication primarily analog. In analog transmissions, the magnitude of the signal is proportional to the quantity being measured or the magnitude of the change commanded of a particular degree of freedom. Ordinary mechanical master-slaves and unilateral manipulators both use analog communication. If developments in space technology indicate trends, analog communication will eventually give way to digital communication, especially where distances are great and where digital computers are added to supplement man or to compress information.* There is a great advantage in having all commands and data expressed in the same format and language.

* Note that teleoperator actuator commands are commonly three-valued; i.e., (1) rotate right, (2) rotate left, (3) do nothing. This fact could lead to trinary rather than the binary coding now common in computers and space communication.

When operator and actuators are physically close, each degree of freedom can be handled economically with a separate communication channel, viz., a metal cable, hardwire, or hydraulic link. As distance increases, multiwire cables are replaced by single strand cables and finally by electromagnetic or perhaps acoustic waves. When this happens, the commands for each degree of freedom and data from all sensors (in short, all information) often share the same communication channel. Sharing is accomplished by time or frequency multiplexing. In time multiplexing, synchronous electrical or mechanical switches sample each sensor periodically. In frequency multiplexing, data from different sensors are impressed upon subcarriers at different frequencies. In space work, time multiplexing is more common.

The act of impressing information upon a communication channel is termed "modulation," and varieties of modulation exist in bewildering confusion. Amplitude and frequency modulation have been employed for decades in industry and scientific telemetry. In space technology, however, pulse modulation seems to be gaining the upper hand, pulse-code modulation (PCM) in particular. Although PCM requires more power and bandwidth than the well-proven and reliable PAM (pulse amplitude modulation), PCM is better matched to the digital computers widely used to interpret, compress, and process large quantities of data. Although space program experience may not dictate future developments in teleoperators, it seems likely that sophisticated teleoperators will draw on this huge reservoir of experience.

Turning back to the basic types of links, we find that two types—the electromagnetic (radio, light) and the acoustic—"broadcast" or "beam" their signals through space or water. In either case, the signals are attenuated by the inverse square law and absorption in the medium. These laws are well known (Krassner, 1964; Machol, 1965). The hardwire, mechanical, hydraulic, and pneumatic links all depend upon a physical "conduit" to convey signals back and forth. The conduit of course absorbs a portion of the signal, but the attenuation of the inverse-square law is circumvented. Noise is usually lower on these links, although there may be cross talk between adjacent hardwire conductors.

One of the critical spots in any physical signal conduit is the spot where it pierces the barrier between the operator and the hostile environment. In hot cells, for example, radioactive dust may leak around and through mechanical manipulators. In a deep-diving submersible, every hull penetration is a weak spot in an environment where pressures are great. For this reason, hull penetrations are nearly always electrical (which are smaller and allow no fluid passage) rather than hydraulic. The basic constraints limiting the use of physical links are cost and the inconvenience of maintaining or dragging vulnerable cables hooked to mobile and distant fixed teleoperators.

Table 4.3 summarizes the characteristics of the basic communication links associated with teleoperator communication.

During its 1968-1969 study of teleoperators applied to satellite repair and maintenance, General Electric examined the characteristics of long-distance radio links between ground-based controllers and satellite-borne teleoperators (General Electric, 1969). The teleoperator satellite, shown in Fig. 2.3, carried two omnidirectional antennas and one high-gain antenna, which linked the spacecraft to NASA's Space Tracking and Data Acquisition Network (STADAN). In concept, commands and telemetry would be transmitted from the STADAN station working the satellite back to the operator along the cables and microwave links comprising NASA's ground communication system (NASCOM). Communication satellites could be used to relay information whenever the teleoperator satellite is beyond the range of STADAN stations. General Electric proposed two uplink channels carrying tracking signals (range and range rate) on one and, on the other, manipulator control and TV camera control signals as well as satellite commands. Three downlink channels were reserved for tracking data, TV signals, force feedback information, and satellite "housekeeping" telemetry. The manipulator control and feedback signals, though basically analog signals, were converted to PCM for transmission.

## THE COMPUTER SUBSYSTEM

When teleoperators are engaged in space and undersea exploration, a general purpose computer will be desirable for such functions as:

1. Data compression and processing.
2. Lengthy computations (i.e., coordinate transformations).
3. Preview and supervisory control (see chapter 5).
4. Data memory in cases where subroutines must be stored.
5. The generation of artificial displays for the operator in situations where visual displays are impossible.
6. Forecasting the outcome of specific operator actions (similar to preview control, only looking into the future rather than guessing the present).

The presence of a general purpose computer in a teleoperator system may not markedly diminish the need for many small, local, analog and digital computers associated with sundry subsystem functions. Most sophisticated teleoperators, for example, will have one or more thermo-

Table 4.3 Characteristics of Teleoperator Communication Links

| Type of link | Characteristics | Examples |
|---|---|---|
| Mechanical | Analog and continuous. One cable or tape per degree of freedom or sensor. Power may be transmitted at the same time as commands. Limited to short ranges (tens of feet). Hard to make good barrier seals. | Mechanical master-slaves. Tongs, ball-in-socket manipulators. Micromanipulators in electronics and biology. Prostheses. |
| Hardwire (electrical) | Analog and/or digital. Continuous or multiplexed. Cables may be many-stranded or multiplexed. Power may be transmitted at the same time as commands. Limited to a few miles in length, except when adaptable to terrestrial communication nets already in existence; i.e., commercial and government hardwire networks. Cables are inconvenient and often vulnerable to the hostile environment. | Electrical master-slaves. All fixed and some mobile unilateral manipulators. Submersible hull penetrations. Undersea stations (Benthic Lab). Some prostheses. |
| Hydraulic | Same as mechanical links. Leakage is a problem. | Heavy-duty unilateral manipulators (forging types) and exoskeletons (Handyman). |
| Pneumatic | Same as mechanical links. Leakage is a problem. | Some prostheses (Heidelberg arm) and special purpose manipulators. |
| Electromagnetic | Analog or digital. Continuous or multiplexed. Length of link unlimited. Inverse-square-law and medium attenuation. Extraneous noise is a problem. | Radio-controlled mobile manipulators (MRMU). Potentially applicable in all space operations. Lasers. |
| Acoustic | Relatively unexplored. Bandwidths more restricted than radio. Undersea problems include high absorption, refraction, scattering, and the presence of multiple paths. | Potentially applicable in all underwater operations. Disney Audio Animatronics System. |

statically controlled regions, some voltage and power regulators, attitude-stabilization devices, and so on. These local control loops, with their small analog computers and/or logic circuits will have little of the flexibility and power of the general purpose computer. They are, however, ubiquitous in most complex machines.

The teleoperator computer is more likely to be digital than analog. Analog computers are very useful in specialized applications, such as autopilots, but do not have the memory capacity and versatility needed for advanced teleoperator concepts. The digital computer fits very nicely into the teleoperator that employs pulse-code-modulated (PCM) communication for commands and sensor data. PCM is the natural "language" of computers and most advanced remote-control systems.

The physical location of the computer depends upon the application. In actuality, there may be two (or even more) computers in a complex teleoperator. On a distant planet a teleoperator will probably require a local general-purpose computer for supervisory control and data compression prior to transmission. The operator on the Earth will want another computer for preview control because of the long time delays involved and for display generation.

The same possibilities occur in undersea exploration save for the time-delay problem. Bandwidths in undersea communications systems are likely to be restricted (especially if an acoustic link is used) and an on-the-spot computer can improve overall performance greatly by compressing sensor data prior to transmission back to the operator.

Many small general purpose digital computers have been constructed for the manned space flight program. Teleoperator computer technology can build directly upon this base.

## THE PROPULSION SUBSYSTEM

Mobility is essential to the success of many teleoperators. Man's incomparable dexterity would be next to useless if he could not walk about and apply it. A teleoperator might employ any form of locomotion that has been invented, however, the pertinent column in Table 4.2 indicates that each application area has concentrated upon only a few types of propulsion. As a generalization, it can be said that teleoperator propulsion tends strongly to be unspecialized because the keynote of the teleoperator is versatility. On land, for example, tracked vehicles are usually preferred to wheels which demand a smooth, unlittered, hard pavement. In a similar vein, buoyant submersibles are usually superior to ocean-bottom crawlers because they can move more freely.

In orbital or interplanetary prime space, the so-called "reaction engine" is the only practical prime mover. The engines required are of course rocket engines, but small ones suffice in this case because only small thrusts are needed for orbital adjustment and rendezvous. Of the two basic types of "chemical" engines—solid and liquid—only the liquid engines have the multiple restart capability and throttleability essential for precision maneuvering.* Even with the choice narrowed this far, there are many propellant combinations to choose from: bipropellants, monopropellants, cold pressurized gas, etc. This selection problem was faced during the Independent Manned Manipulator (IMM) study carried out by Ling-Temco-Vought (LTV) and ANL for the Marshall Space Flight Center in 1966 (Ling-Temco-Vought, 1966). Its approach and conclusions are most useful here.

Table 4.4 Guidelines and Requirements for Propulsion Subsystem Design for the Maneuvering Work Platform (MWP)

| | | |
|---|---|---|
| Guidelines | In-orbit service and maintenance | |
| | A single-point failure will not prevent a return to the parent ship | |
| | Minimum exhaust-plume effects (heating, etc.) | |
| | Maximum use of existing hardware | |
| | Expendable resupply at 120-day intervals | |
| | Re-service of vehicle following each mission | |
| Requirements | Required impulse per task | 45,000 lb-sec |
| | Total yearly impulse | 739,000 lb-sec |
| | Tasks per year | 62 |
| | Number of thrusters | 24 |
| | Thruster thrust | 13 lb |

Two vehicles were examined during this study: a Maneuvering Work Platform (MWP) and a "Space Taxi." Both vehicles had electrical master-slaves attached but the MWP will be used here as a reference design. The MWP guidelines and requirements are listed in Table 4.4, and are representative of orbital space teleoperators circa 1970.

During the LTV propulsion study, one bipropellant (nitrogen tetroxide and Aerozine 50), two monopropellants (hydrazine, 90 percent hydrogen peroxide), and cold nitrogen gas propellant were investigated in detail. The bipropellant combination is used in the Apollo Program and is a good representative of the state of the art. It does, though, have a high combustion temperature which leads to exhaust-plume heating problems. Hydrogen peroxide also has been used extensively in space, but is not easily stored for long periods. Cold nitrogen gas under pressure is in-

---

* Electrical propulsion might prove desirable in more advanced teleoperators.

nocuous enough, but it possesses no intrinsic energy and consequently has a very low specific impulse. The best choice for the MWP was reported to be hydrazine.

The General Electric teleoperator satellite also postulated a hydrazine propulsion subsystem for rendezvous, maneuvering, docking, and stabilization. Here, the propulsion system consisted of two 26-pound thrust rendezvous engines, eight 2-pound thrust, and sixteen 0.5-pound thrust engines for the other propulsive functions.

Mobility beneath the sea involves remarkably similar considerations. A small, manipulator-carrying submersible hovers when it possesses neutral buoyancy in much the way a satellite "floats" in space. As a submersible approaches its target, it must maneuver and dock, just like its space counterpart. The "engines" in this environment are nearly always propellers or water jets that can be controlled in thrust level, thrust direction, or both. During actual manipulation tasks, the submersible will generally be anchored to the target and attitude changes can be made with the docking arms or "grapplers"; propulsion is needed only during approach and docking.

The small submersibles require speed of only a few knots. The *Autec I* vehicle can cruise at 2 knots for 8 hours, and has a maximum submerged speed of 3 knots (North American Aviation, 1966). It is designed to hover $\pm$ 5 feet at depths below 200 feet. The Deep Submergence Rescue Vehicle (*DSRV*) is a couple of knots faster and must be able to hover over a given spot against a 1 knot current, at attitudes up to 45° from the horizontal. Most submersibles meet such requirements through the use of screws of various types.

Usually a single main screw provides propulsion until target rendezvous begins. Then auxiliary screws or jets mounted in pairs around the vehicle provide precise control for hovering, up-and-down motion, and any other maneuvers needed.

Tracked vehicles may have important applications on those portions of the continental shelves where bottom conditions are suitable. The Scripps Remote Underwater Manipulator (RUM) is a major example of bottom crawlers. RUM was propelled through two independently driven electric motors (Anderson, 1960). Each track was powered by a 7½ horsepower, 800 rpm, dc motor. Power was provided from shore in all RUM tests through a 5-mile-long multiconductor cable.

The nuclear industry—first to use manipulators on a wide scale—was also the first to place them on vehicles. Most AEC laboratories have developed their own or purchased commercially made mobile manipulators for emergency use. The PaR-1 vehicle (Fig. 2.5) is representative of the smaller tracked vehicles in this class. Mobot and MRMU illustrate the medium and large classes respectively (although they are both inoperative at present). Most manipulator-bearing vehicles used in AEC facilities are driven by electric motors and depend upon long power cables. MRMU, which is radio-controlled, is an exception; it is powered by a gasoline engine and can attain a speed of 35 mph. MRMU's chassis is a converted, full-tracked Army XM474 cargo carrier. The many manipulator-carrying vehicles now in use in different nuclear installations are described by Homer (Homer, 1966).

In many nuclear operations, the working areas are usually quite cluttered (especially during emergencies and rescue operations) and therefore unsuitable for wheeled vehicles on the floor. Manipulators mounted on wheels riding on overhead crane-type tracks see considerable service in such situations. The large wall-mounted manipulators installed in the E-MAD building at the AEC-NASA Nuclear Rocket Development Station in Nevada operate on this principle (Fig. 4.1). Driven by electric motors that pick power off metal strips along the E-MAD walls, these manipulators can range up and down the length of an immense hot cell.

The only major type of vehicle not mentioned so far in this section is the large, heavy-duty forging manipulator that transports hot forgings and billets in foundries. Teleoperators employed in mining and construction work would be similar in size and power, but would undoubtedly substitute tracks for wheels.

## THE POWER SUBSYSTEM

When motion is communicated between the operator and the actuators by mechanical means—cables, metal tapes, etc.—the power source is usually man himself, as in most prosthetic devices. The human is a good power source when the target is close by, not too heavy, and the tasks are not too tedious.

If commercial power lines are nearby, the power problem is minimized. In many hostile and distant environments, however, neither man's power nor commercial electricity can be conveniently communicated to the actuators. Teleoperators then can either carry power sources along with them or try to extract energy from the environment.

Except for a few space concepts employing solar cells and teleoperators used near commercial electrical power, transportable power sources are dominant. Chemical sources, such as internal combustion engines, trail far behind human power in current hardware. Batteries and compressed gas bottles provide limited amounts of power, especially in prosthetics. Nuclear power plants seem promising for future deep space and undersea activities. Table 4.5 gives specifics by application area.

Figure 4.1  The General Electric electric unilateral boom joints on the Wall-Mounted Handling System, at the E-MAD Building, NRDS, Nevada. Photo was taken before the concrete shielding walls were poured. (Courtesy of General Electric Co.)

Manipulation requires that raw power—heat, electricity, sunlight, and so on—be converted into mechanical energy. As subsystems are defined here, the task of converting raw power to mechanical energy falls to the transducers in the actuator subsystem; that is, to the electrical and/or hydraulic motors, pistons, etc. (These transducers will be covered in detail in the next chapter.) The actuator subsystem usually consumes power in a form different from the raw power produced by the power subsystem. The same is true with the other subsystems, except that they are more likely to require electricity than hydraulic power. Electricity, after all, is the life's blood of modern man-machine systems. In most cases, therefore, the teleoperator power subsystem will need a rather elaborate power conversion section that converts the basic power generated by the source into power for each subsystem at the correct voltage, pressure, and degree of regulation needed.

There is a slow unmistakable trend in power subsystem design toward direct conversion devices, such as fuel cells and thermoelectric elements. It is tempting to say that the removal of moving parts can only improve reliability, but teleoperators are not ordinary machines. Teleoperators, for example, *must* have many moving parts if they are to succeed. Reliability may actually be improved by generating mechanical motion in the power subsystem directly and then conveying it more or less directly to the actuators, treads, wheels, and other moving parts. An automobile's hydraulic power take-off is a good example. To summarize this rather

Table 4.5 Teleoperator Power Requirements and Power Plant Perspective

| Application area | Teleoperator (only) power range (kw)[1] | Commercial electric lines | Solar | Human power | Chemical sources | Nuclear sources[4] | Compressed gas sources |
|---|---|---|---|---|---|---|---|
| Aerospace | | | | | | | |
| Orbital vehicles | 0.1–5 | | P | P | P* | P | |
| Planetary probes | 0.5–2 | | P | | | P | |
| Space chambers | 0.1–0.5 | P | | U | | | |
| Undersea | | | | | | | |
| Submersibles | 0.5–100 | U | | | U | U[3] | |
| Crawlers | 5–500 | P | | | P | P | |
| Laboratories | 0.1–1 | | | | P | P | |
| Nuclear | | | | | | | |
| Master-slaves | 0.1–0.5 | U | | U | | | |
| Vehicles | 1–500 | U | | | U | | |
| Unilateral | 0.1–1 | U | | | | | |
| Terrestrial transportation | | | | | | | |
| Walking machines | 10–500 | | | | P | | |
| Man-amplifiers | 50–250 | | | | P* | | |
| Medical | | | | | | | |
| Prosthetics | 0.1–0.5 | | | U[2] | U* | P | |
| Surgery | ? | P | | | | | U* |

64

Table 4.5 Teleoperator Power Requirements and Power Plant Perspective (Continued)

| Application area | Teleoperator (only) power range (kw)[1] | Commercial electric lines | Solar | Human power | Chemical sources | Nuclear sources[4] | Compressed gas sources |
|---|---|---|---|---|---|---|---|
| Chemical<br>Master-slaves | 0.1–0.5 | | | U | | | |
| Metal processing<br>Forging manipulators | 10–500 | U | | | U | | |
| Electronics<br>Master-slaves | 0.1–0.5 | P | | U | | | |
| Construction<br>Vehicles | 10–500 | | | | P | | |
| Public service<br>Vehicles | 10–500 | | | | P | | |
| Entertainment<br>"Puppets" | 0.01–1 | U | | | | | |

Key to symbols:
P = proposed.
U = in use now or in past.
* = discussed in text.

[1] Average requirements; peak demands may be much higher; includes lighting, communications, etc.
[2] Human motions (breathing, etc.) may be transformed into electrical energy and stored for later use.
[3] Classified information.
[4] Nuclear reactor or radioisotope heat sources.

elusive point: teleoperators always have moving parts, and static power conversion may be less reliable than dynamic power conversion, i.e., turbogenerators, etc.

It is impractical to survey all power sources used on or proposed for teleoperators. Some entries in Table 4.5 are well-developed, e.g., gasoline engines (a type of "chemical" engine). A few others are far enough along in development to be used as examples.

One is accustomed to thinking in terms of solar cells for the power source on long, unmanned trips to the planets; but, for short, manned missions in orbital space, chemical power sources are usually superior on a weight/cost basis.

The Maneuvering Work Platform (MWP) examined by LTV and ANL assumed a one-year operational life, with approximately 62 eight-hour missions during that period. Since the safety of the astronaut-operator was paramount, no single-point failure modes were permitted. An excellent view of typical teleoperator power requirements in orbital flight can be found in Tables 4.6 and 4.7. In making the power estimates, the specific teleoperator task was assumed to be the erection of a space telescope. On this mission the average electrical power required was about 250 watts.

For forays of a few hours duration from a parent satellite, the only power sources that proved reasonable on the bases of weight and volume for the MWP were chemical power-plants that could be recharged upon return to the parent ship. Batteries, fuel cells, and chemical turbogenerators all met the basic requirements. Chemical turbogenerators, however, required large quantities of reactants and produced severe heat loads on the MWP environment-control subsystem. Fuel cells with the life expectancy and cyclic capability required for the MWP missions

Table 4.6 MWP Electrical and Electronic Equipment Power Requirements[a]

| Equipment | Average power required |
|---|---|
| 1. Communications | 13.0 watts |
| 2. Radar | 50.0 |
| 3. Displays | 8.0 |
| 4. Control electronics | 4.0 |
| 5. Stability and control electronics | 36.0 |
| 6. Environment-control subsystem | 68.0 |
| 7. Thrusters | 30 |
| 8. Grapplers (docking and anchoring) | 124 |
| 9. Floodlight | 80.0 |
| 10. Hand tool (250 w; 10% duty cycle) | 25.0 |

[a] Ling-Temco-Vought, 1966.

Table 4.7 MWP Electrical Energy Requirements Analysis

| Mission phase | Equipment operating during mission phase[a] | Energy requirement (watt-hours) |
|---|---|---|
| Orbital transfer | Items (1) (2) (3) (4) (5), and (6) | 107.4 ± 11.2 |
| Docking and unstowage of cargo | Items (1) (3) (4) (5) (6) (8), and (9) | 285.8 ± 30.7 |
| EVA erection of telescope | Items (1) (6) (9), and (10) | 720.9 ± 85.6 |
| Orbital transfer and maintenance trip | Items (1) (2) (3) (4) (5) (6), and (9) | 204.7 ± 13.6 |
| Maintenance (mission worksite) | Items (1) (6), and (9) | 238.8 ± 31.5 |
| Intermittent operation during mission | Items (7) and (8) | 202.6 ± 33.4 |
| | Total | 1760.2 ± 266.0 |

[a] See entries on Table 4.6 for number key.

presented too many development problems; therefore, fuel cells were not considered for the 1970 time period. The only safe choice left was the electric battery. Of the several possibilities, the silver-cadmium cell was considered most likely to meet the operational life and deep-discharge requirements.

Significantly, an examination of power requirements for a larger, farther-in-the-future (1975) Space Taxi led LTV to the choice of fuel cells. It was presumed that the fuel-cell development problems would be solved by 1975.

The power-plant considerations for small submersibles run almost parallel to those for orbital vehicles. In both application areas, relatively short expeditions from a mother ship are common. Under the sea, though, the power requirements are larger, usually because the entire vehicle is larger and the power subsystem must drive the propulsion system. Except for the nuclear power plant on the *DSRV-1*, small submersibles will use batteries for the present and the near future, with the fuel cells becoming more interesting in the 1970's. (In underwater oil well operations shore electrical power may be available at the working site.)

On dry land, electrical power sources are the most popular choice of teleoperators, except, of course, for human-powered master-slaves and artificial limbs. Nuclear and solar power are not significant today in terrestrial teleoperators. Where commercial electric lines are not available or impracticable, the only extant power sources are those that utilize chemical reactions and, in some prostheses, the energy of compressed gas.

One of the more intriguing terrestrial applications of teleoperator principles is the man-amplifier. The best-publicized conceptual engineering efforts along this line are the Cornell Aeronautical Laboratory man-amplifier studies sponsored by the Department of Defense, and the more recent "Hardiman" concept being investigated jointly by the U.S. Army and U.S. Navy. In all man-amplifiers, the human operator wears an exoskeleton with which he can perform superhuman tasks, such as lifting ton-size weights.

Superhuman performance manifestly demands superhuman power subsystems. A strong man can develop a horsepower or two for a few seconds. To be worthwhile a man-amplifier should have tens of horsepower over spans of several hours. To be transportable the power supply is likely to draw on chemical energy.

The Cornell studies (Clark, 1962) in the early 1960's gave us the first estimates of power requirements for a man-amplifier. About 10 horsepower was estimated for the Cornell concept. More recent Hardiman power estimates are higher: 15 horsepower-plus just for standing still and about three times that for walking. Evidently first-generation man-amplifiers will consume as much power as a small automobile.

To power their man-amplifier, the Cornell Aeronautical Laboratory proposed two systems:

1. A hot-gas-powered electrohydraulic system.
2. A hot-gas power system in which the actuators would be powered by the hot gas directly.

Neither of these power supplies was investigated in detail, either on paper or in the laboratory, but each would undoubtedly be rather bulky. Both the hot radiator in the hot-gas-hydraulic system and the hot motor exhaust in the direct-power system would be hazardous. As we shall see from the next example, present concepts for man-amplifier power supplies are large and undeveloped in comparison with power sources employed in the prosthetics field.

Most contemporary artificial limbs and orthotic devices are moved by human muscles. When this is impossible or awkward, a small power source generating a few watts may prove a blessing to a handicapped person. Unfortunately, little research has gone into what the medical people call "external power supplies" (U.S. Government, 1966). The only power sources that have been investigated in any depth are compressed carbon-dioxide cylinders and electric batteries. Hydrogen peroxide is occasionally mentioned in the medical literature, but it has not been explored in terms of hardware.

The space program contributes directly to the prosthetics field through its efforts to develop long-life, rechargeable, sealed batteries of minimum weight (Szego, 1966). The best battery for prosthetics use today is probably the nickel-cadmium cell, a power source used on many unmanned satellites. The lighter-weight, silver-zinc cell is coming into operational use in space and will probably be found powering artificial limbs before long.

Although batteries can be recharged conveniently and compare well with compressed-gas power sources, the latter have gained ascendancy in the prosthetics field. Mainly, this is because $CO_2$ capsules came into use before lightweight, reliable, sealed batteries were available and $CO_2$ actuators are simpler and lighter than their electrical counterparts. In addition, electric prostheses have not been notably successful. One problem is the whine of high-speed electric motors. Electrically powered artificial limbs are easy to control, however, and are more easily integrated with electromyographic and other electrical control schemes.

Compressed $CO_2$ is energetic enough to power artificial arms and hands for satisfactory periods of time (U.S. Government, 1966). Standardized steel capsules can be refilled with liquid $CO_2$ by the prosthesis wearer himself. The capsules are small enough to be concealed under the clothing in many instances. In current practice, $CO_2$ pressure is reduced by a regulating valve to 100 psi or less and conveyed directly to the servo valve controlling the artificial limb. $CO_2$ cylinders are common because they have proven simple, reliable, safe, and convenient to use.

Although teleoperator power subsystems now in existence rely heavily upon chemical sources of energy and electric power lines, nuclear and solar power subsystems will certainly be developed for future space and undersea exploration. Changeovers from chemical to nuclear power will come first on those missions where power is needed over long periods of time and where resupply with chemical fuel is impossible or too costly.

## THE ATTITUDE-CONTROL SUBSYSTEM

The presence of an attitude-control subsystem on a teleoperator presumes that some portion of the teleoperator is free to rotate with respect to the target or some set of reference axes. On terra firma, teleoperators generally do not need attitude-control devices because they are either fixed (master-slaves attached to a hot-cell wall) or vehicles with essentially fixed attitudes (MRMU). An attitude-control subsystem has no place on such teleoperators, unless they happen to employ a walking mechanism for translation. Some walking machines, particularly the two-legged type, do change the attitude of the operator during the walking cycle. In these cases, attitude control becomes a matter of balance and the elimination of attitude control is likely to disorient the operator. This is more in the province of actuator design, since, in the end, it is the actuators (legs) that balance the machine and stabilize the motion of the cockpit and operator. (Chapter 7 discusses these matters in more detail.)

Attitude control becomes critical on "hovering" submersibles and spacecraft which must attain and maintain certain attitudes with respect to sunken submarines, space telescopes, or other targets.

There are three important ways to control the attitude of a vehicle that is free to rotate in one or more degrees of freedom: (1) reaction engines (jets or screws), (2) gyros, and (3) docking arms or manipulator arms that can exchange angular momentum with the target or some other object.

The LTV–ANL–MSFC Independent Manned Manipulator study again gives us a reference point. Considering the maneuvering and docking required during the erection of an orbital telescope, the study produced angular acceleration requirements of 8 to 30 degrees/sec$^2$ on the pitch roll, and yaw axes. Given the size and mass of the Maneuvering Work Platform (MWP), these requirements were translated into moment and angular-momentum requirements. To meet these requirements an all-jet reaction system was compared with a hybrid jet-gyro system. On the basis of weight and volume (including allowances for extra electrical power drain), the former was selected for the MWP.

Under the ocean, attitude-control requirements are qualitatively similar to those in space, but quantitatively different because of the larger vehicle sizes, turbulence, ocean currents, and the viscosity of seawater. Attitude control is aided in deepsea work by (1) the ready availability of propellant (water); (2) the presence of a strong gravity field that permits attitude trimming by shifting the center of mass relative to the center of buoyancy (say, through the use of pumped mercury), and (3) the use of anchors.

Small submersibles may use translation propulsion systems for attitude control. The main propulsion system, however, may not prove suitable in maneuvers necessitating frequent propeller reversals. For this reason, special nozzles and/or ducted propellers (called "cross-hull thrusters") usually are located around the hulls of manipulator-carrying submersibles.

The attitude of a submersible or spacecraft is so easily perturbed that operating philosophy recommends stabilizing the vehicle with respect to the target with grappling arms that mechanically or magnetically "grab" the target structure and position the vehicle relative to it. Precision attitude control, then, is a function needed only when the vehicle is approaching and leaving the target or mother ship.

## THE ENVIRONMENT-CONTROL SUBSYSTEM

Environment control, like attitude control, becomes critical in teleoperator design when outer space, the undersea, or radiation fields, are invaded. Teleoperators in strong nuclear radiation fields have to be shielded from the deleterious effects. A more difficult problem—temperature control—is important in outer space where there is no atmosphere or ocean to keep the power-consuming and (consequently) heat-producing teleoperator cool.

The two problems are not completely independent. On unmanned missions, such as the Benthic Laboratory or a Martian probe, no life support equipment may be needed, but the artificial atmosphere that could serve as a heat sink for a variety of electronic gear will also be missing. In such cases, adequate heat conduction and/or convection paths must be provided to an external surface where the heat can be removed by radiation to space or conduction to seawater. Of course, the existence of a life-support system does not eliminate the problem of thermal control, it just transfers it to one of cooling the artificial atmosphere. The artificial atmosphere may not be sufficient or convenient for cooling, say, the auxiliary power unit, and special coolant loops will have to be provided.

On short, manned space missions, the environment-control subsystem must: (1) provide oxygen, (2) remove carbon dioxide, and (3) remove heat. For a relatively short mission, with resupply of expendables from a parent ship, the design of the environment-control subsystem is simplified in the following ways:

1. Bottled oxygen can be used instead of regenerative equipment.

2. Atmospheric contaminants do not have time to become concentrated, and only $CO_2$ needs to be removed.

3. Heat rejection can take place through a sublimator/evaporator rather than a radiator, which expends no materials but is heavier and occupies more volume.

Undersea manipulator-carrying vehicles have similar missions in terms of time and environment-control requirements. The major difference is the replacement of the external vacuum environment by cold seawater. Many of the principles used in designing space environment-control subsystems also apply to submersibles. There is now an immense body of literature dealing with life support in various hostile environments (Vinograd, 1966; Webb, 1964).

## THE STRUCTURE SUBSYSTEM

The teleoperator structure performs one or both of two functions: (1) encapsulation and protection of the operators, and (2) service as a framework to support attached teleoperator components.

When protecting the operator, the structure subsystem becomes essentially a pressure shell. At great depths in the ocean, this shell may be a major design problem. Both in space and under the ocean, operator capsules tend toward spherical and ellipsoidal shapes.

A mere platform suffices for the human operator in a terrestrial walking machine. In mechanical master-slave manipulators, all structural support is provided by a simple horizontal tube penetrating the barrier separating the operator from the hostile environment; the master and slave ends of the teleoperator hang from this tube. Vehicles such as MRMU are not markedly different structurally from an ordinary truck, bulldozer, or tank. In short, few generalizations can be made about teleoperator structures. Each one is built to meet the needs at hand.

# V
# THE CONTROL SUBSYSTEM

To control his machines man acts primarily as a goal-setter and an error corrector. He decides what he wants the machine to do; he plans the strategy; he gages the machine's deviation from desired performance; and he manipulates the machine's controls in a way that reduces the error. He does this when he steers a car along a winding road and when he picks up a sample of lunar soil with a teleoperator hand from a distance of a quarter-million miles. These words oversimplify the situation. Any control system that counts a human being among its elements is complex and difficult to describe scientifically because man himself is so complex and difficult to describe.

Why, then, admit man to the teleoperator control loop? Machines can certainly detect their own errors and correct them; autopilots and home heating plants do this very nicely. The reason for man's presence stems from his ability to set strategy and to deal with the unexpected —those situations we cannot preprogram into a machine's memory. Man is an adaptive creature; and, if teleoperators are to be the extension of man, they must be adaptive also. To illustrate: Could a pure machine uncomplicated by man's presence figure out how to repair a ruptured oil pipeline far out on the continental shelf?

In principle, the answer to the foregoing question is "yes." Adaptive machines, machines that learn from experience, can and have been built. They are true robots. Today's robots, however, cannot approach man's adaptability, versatility, and intelligence. It would take many ruptured pipelines before a robot learned how to fix them. For decades, at least, man-operated teleoperators will reign supreme in those hazardous and distant spots where man prefers to send machine proxies.

Most extant teleoperators are "pure" man-machine systems; that is, man is *always* in the control loop. As teleoperator technology progresses, though, preprogrammed subroutines are being added to relieve the operator of those wearisome, repetitive tasks that can be done better by machines. A very simple and basic preprogrammed subroutine is one which stops teleoperator arm motion when limit switches indicate that self-inflicted damage is imminent. Most complex machines include similar localized reflex control arcs that intrinsically react faster than man. Subroutines are also extremely useful in space operations—say, lunar exploration—where there is significant signal time delay between operator and teleoperator hands. Such subroutines, which are intrinsic to Sheridan's and Ferrell's *supervisory control* approach (Sheridan, 1963) do not add to a teleoperator's intelligence or adaptability, but they improve overall effectiveness considerably, especially where time delays are large. In principle, then, a continuous spectrum of teleoperators exists between the pure, man-always-in-the-loop extreme to the completely preprogrammed, dexterous, general-purpose machine possessing only an ON-OFF switch, in other words, a *robot*. As technology progresses, we may expect to see teleoperators move toward the robot end of the spectrum.

The ingenuity of man and his passion for making machines that emulate himself should not be discounted. The future may soon see the addition of adaptive or artificially intelligent subroutines to teleoperators. At first, some of the simpler, more routine decisions might be machine-made. Eventually, both preprogrammed and adaptive subroutines might be added until man could say to a machine, "Go and explore the galaxy for me." Philosophically speaking, the teleoperator may be a transitional man-machine system that presages generations of machines that are man-like, man-directed, man-serving, and yet self sufficient save for a few spoken commands from their masters.

## MAKEUP OF THE CONTROL SUBSYSTEM

At the core of the control subsystem is the human operator. Toward him flow feedback data that describe the positions and velocities of the teleoperator's hands, arms, and other actuators as well as the objects being manipulated. From him flow the commands that will reduce (hopefully) the error he perceives in teleoperator performance. The human brain is the goal-setter and the error computer, planner, and decision maker, although a computer may supplement some brain functions.

Two man-machine interfaces are of paramount importance. First, feedback information from the machine part of the teleoperator must be "read into" the brain so that a performance error can be computed. Current terminology calls the device that translates machine sensor readings into signals comprehensible to the brain a *display*. A display may be simply a faithful television view of the scene or it may be a symbolic display, such as a meter indicating the grip force exerted by the teleoperator hand. The second critical interface separates man from the teleoperator actuators, as well as other teleoperator subsystems under the operator's direct control. Man's commands to his machine partner stream through his central nervous system to his arms, hands, eyes, tongue, and other parts of his anatomy that can create mechanical, sonic, and electrical signals. These signals cross the man-machine interface and activate *controls* that convert them into commands comprehensible to the rest of the teleoperator.

The complete circuit from man to machine and back to man is the *control loop*. Information courses around this loop, which may be augmented by computers here and there. The successful operation of the teleoperator depends upon the successful encoding, transmission, and translation of this data stream.

## MAN AS AN ELEMENT IN THE CONTROL SUBSYSTEM

The human operator eludes precise definition. If he did not, control engineers could formulate an elegant *human transfer function or human describing function* that would mathematically describe what man would do when confronted with feedback data and decisions to make. The human transfer function describes what a normal man will do given a specific input. In the next chapter, we will describe some of the human transfer functions that have been synthesized for extremely limited situations. Unfortunately, they have scant utility in teleoperator control theory, except for helping predict system stability and in very special situations. In teleoperators as nowhere else, man is an adaptable, rather unpredictable element that cannot be encompassed by formulas.

In lieu of precise mathematical human describing functions, words must suffice. It is common to describe man in terms of his *input-output* characteristics, just as if he were an electronic control component or *black box*. The words, however, can only guide us in the design of the teleoperator control subsystem.

The sensory input channels leading to man's brain are many. We know how to use them but not how or why they work as they do. From this wide selection, only four of our senses are in actual use today in teleoperator work; vision, audition, and the cutaneous and kinesthetic senses; i.e., sight, sound, touch, and the sense of position and motion.

Sight is by far the broadest channel carrying feedback information to the operator. In fact, it is the *only* input channel in most teleoperator systems. This is true because sight is practically indispensable* in manipulatory tasks—we *have* to have it—and visual channels are relatively easy to build (windows, TV etc.). Force feedback is present in mechanical and electrical master-slaves as well as some walking machines and man amplifiers currently under development. Proprioceptive feedback or sense of limb position can be achieved by using exoskeletal controls that maintain the same configuration as the actuators. The man amplifier possesses such exoskeletal controls. Touch sensation, as opposed to gross force feedback, is highly desirable in a teleoperator but sometimes not worth the cost of instrumentation; it has not been developed to the point where it it used regularly. Sound waves coming from manipulatory processes carry alarm or warning signals (viz., a dropped object); and for this reason a few manipulators incorporate microphones.

Despite the paucity of feedback channels in contemporary teleoperators, designers always have as their ultimate goal the faithful reproduction (occasionally, amplification) of most of the sensations that would normally be experienced by an unaided human actually doing the job of the teleoperator. In practice, they settle for much less. Of course, no one reproduces all aspects of a hazardous environment for the operator—just those aspects of the environment that will aid manipulation. For example, the forces experienced by the machine body of a man amplifier would crush the human operator if they were not attenuated.

Humans also have subtle input problems. For example, the all-important visual channel is subject to a great variety of optical illusions and signal disortion. Then, there is operator fatigue which can seriously distort the information presented to the decision-making and command-generating portion of the brain. Fatigue also lengthens the operator's reaction time. Finally, man's senses are far from the easy-to-analyze linear transducers that engineers like so much; that is, the intensity of a stimulus perceived by the operator is not proportional to the actual physical magnitude of the stimulus. Instead, each sensory channel seems to exhibit a different power law relationship.

To illustrate the complexity of the problem, some evidence suggests that, if a system has anthropomorphic features, the operator instinctively employs his long-used anthropomorphic responses. This may be undesirable if the task or feedback is nonanthropomorphic. Yet, in hot-cell work the roughly anthropomorphic master-slaves have proven to be highly effective.

* Manipulation by force feedback alone is possible but it is generally not efficient.

Man as an Element in the Control Subsystem 77

**Figure 5.1** An orthotic arm controlled by electromyographic (EMG) signals generated by the amputee's muscles. (Courtesy of Case Western Reserve.)

Human weaknesses are counterbalanced by unexpected strengths that transcend the usual adaptive and integrative powers. Airplane pilots, astronauts, and other operators of complex machines show a surprising ability to handle nonanthropomorphic displays and manipulate controls that certainly seem "unnatural." In fact, man *may* overpamper himself and unnecessarily restrict the machine by making his teleoperators too much after the human mold; although some engineers object to this contention.

Once the human operator has digested the stream of input information and decided upon a course of action, he "emits" a train of command or output signals. Precisely what transpires between input and output in the human transducer has been argued by speculative philosophers for centuries. In other words, we really have little idea of how information is processed in the brain; and for practical purposes we do not really need to know.

To translate his commands into machine language, the operator has at his disposal his hands, feet, head, in fact any part of his body that moves, even his eyeballs. By far the most useful output channel depends upon the motion of the human hands. In current teleoperator design, the preponderance of hand-actuated controls is even more marked because manipulators are really machine analogs of man; and it seems eminently logical to control hands with hands. Similarly, in a biped walking machine it is natural to control legs with legs.

When the teleoperator must be steered or flown, or it possesses more degrees of freedom than the operator can handle with his hands and feet, the human voice may serve as an output channel. Even today, machines can be designed to recognize a small array of spoken commands, such as "turn left" or "stop."

Suppose a handicapped person has no hands or arms to control his artificial limbs. Muscle-bulge switches and shoe switches are sometimes employed. More often, limb remnants and shoulder muscles activate prostheses. A promising human output channel, still in the research and development stage, translates the weak electrical signals created within the body by muscle action into electrical commands a machine can understand. Muscle action potentials (MAPs) form the basis of electromyographic (EMG) control of artificial limbs as well as other types of teleoperators (Fig. 5.1).

## SOME SPECIAL TELEOPERATOR CONTROL PROBLEMS

Most treatises on the human control of machines—*manual control*, as the discipline is often called—deal with aircraft, terrestrial vehicles, and other machines with far fewer human characteristics than teleoperators. Because teleoperators simulate human traits, one would expect that matching the man and machine portions would be easy, but it is actually a most difficult task. Here, we merely list some of the more troublesome aspects of man-machine integration to illustrate how teleoperator control is different.

1. The operator is often located at a point far removed from the mechanical arms and hands he is controlling. In most terrestrial hotcells, where good visual displays and force feedback exist, it is not too difficult for the operator to project himself into the working area; that is, identify his movement with those of the distant hands and arms. The problem here is the provision of good feedback in more difficult applications, such as undersea manipulation.

2. Great distances between the operator and the actuator subsystem introduce signal time delays that confuse the operator. This problem is serious in some space applications; viz., the round-trip signal transmission time to the Moon is about 2.6 sec.

Table 5.1  Some Definitions Used in Teleoperator Control

| | |
|---|---|
| Open loop | No feedback of any kind to operator |
| Closed loop | Some kind of feedback is present. Psychologists call a loop "closed" when vision is present, but engineers usually reserve the term for nonvisual feedback. |
| Preprogrammed | Commands are prerecorded |
| Adaptive | Capable of making decisions based on past experience. |
| Robot | An adaptive machine that needs no human operator, usually humanoid in form. |
| Time delay | Command and feedback delay due to: (1) signal transmission line; (2) coding delay; (3) passive-process delay (inertial effects); and (4) human reaction delay. |
| Preview control | Use of predictive displays (with time extrapolation) to help overcome the effects of time delay. |
| Supervisory control | Use of computers at the operator end to aid decision making and at the actuator end for adaptive control and application of subroutines. |
| Spatial correspondence | Actuators mimic motion of controls (used primarily to describe master-slaves and slaved TV systems). |
| Visual correspondence | Visual display slaved to position of operator's head. |
| Degree of freedom | A dimension of motion in a teleoperator; viz., wrist rotation and elbow pivot. |
| Anthropomorphic | Actuators or controls resemble human body segments in terms of degrees of freedom and how they are articulated. |
| Quickening | The use of time derivatives of teleoperator motion to help the operator predict actuator position and compensate for time delay. (A distant cousin of preview control.) |
| Unilateral teleoperator | A teleoperator in which force and motion can be transmitted only from the operator controls to the actuators. |
| Bilateral teleoperator | A teleoperator in which force and motion can be transmitted from the operator controls to the actuators and vice versa; i.e., the slave arm can move the master arm. (Note: "bilateral" does not imply physical symmetry here as it does in biology.) |
| Rectilinear teleoperator | A teleoperator possessing several degrees of freedom in rectangular coordinates. Generally, these degrees of freedom are associated with over-head bridge-crane positioning systems. "Rectilinear" is often used incorrectly as a synonym for "unilateral." Joints with angular freedom are often termed "polar" in the literature. |
| Master-slave teleoperator | A teleoperator in which forces and torques are proportionally reproduced from the controls (master) to the actuators (slave). A master-slave is bilateral in at least seven degrees of freedom in each arm/hand. All degrees of freedom can be controlled naturally and simultaneously. This term was originated at Argonne National Laboratory. |

80     The Control Subsystem

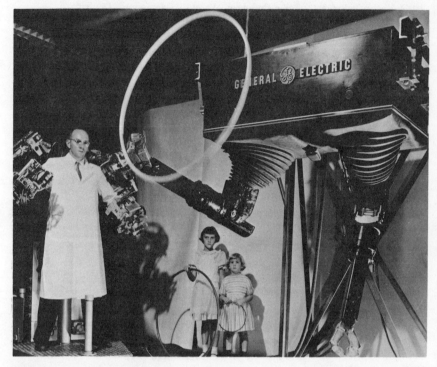

**Figure 5.2**  The General Electric Handyman is a bilateral electrohydraulic master-slave. Built for the Aircraft Nuclear Propulsion (ANP) Program, it is shown here twirling a hula-hoop to demonstrate the degree of coordination possible between the master and slave arms in a bilateral manipulator. (Courtesy of R. S. Mosher, General Electric Co.)

3. A sophisticated teleoperator has so many degress of freedom (over a dozen in many instances) that an operator is hard put to control them in concert unless both controls and actuators possess some anthropomorphic characteristics (Fig. 5.2).

4. If the motions and dimensions of the mechanical hands and arms do not correspond rather closely to the motions of the controls, operator confusion may result. For example, moving a control left should cause the appropriate actuator to move left. (See Table 5.1 for definitions of spatial and visual correspondence.)

PERFORMANCE FACTORS

A cornerstone of systems analysis is the formulation of an overall figure of merit that describes the performance of weapons systems and other complex man-machine conglomerates in terms of a single parameter. The parameter "cost effectiveness" has achieved fame and some notoriety in many fields. Teleoperators have no such advantage; perhaps they are more subtle than weapons systems.

In experiments with manipulators, notably at the U.S. Air Force's Aerospace Medical Research Laboratory, the time taken for a skilled operator to perform a manipulative task has been used as a gage of merit. While useful in comparing different brands of manipulators, this parameter can hardly be expressed in terms of engineering design variables, such as number of degrees of freedom or speed of joint rotation. Teleoperator designers usually rely upon a group of secondary figures of merit, which are collectively optimized by experience rather than systems analysis. We now list those secondary figures of merit related to the control subsystem.

| Figure of Merit | Definitions, Comments, and Intercomparisons |
| --- | --- |
| Torque, force, or grip | Applied to rotating joints and teleoperator hands. The control subsystem should be able to apply force and torque continuously or in graduated steps in response to the controls. Force multiplication between operator and actuator may be desired. Design levels depend upon task at hand. |
| Speed | The linear or angular rate at which a joint moves. Related to torque, force, and the mass of the mechanical hands, arms, and legs. Speed should be controllable in many applications. |
| Accuracy | An arm or hand is accurate if it responds to a command (say, rotate 30° clockwise) with some agreed-upon degree of precision. Precise motion requires good controls. |
| Ease of indexing | The ability of teleoperator appendages to move into prescribed configuration. Computer subroutines are sometimes used to index a teleoperator. |
| Articulateness | A measure of the number of joints and degrees of freedom. Each degree of freedom complicates the control subsystem. |

| | |
|---|---|
| Stiffness | A synonym for teleoperator rigidity. This is a desirable quality (see *sponginess*) |
| Friction | Energy dissipation during motion. This can tire the operator as well as degrade force feedback. |
| Inertia | A measure of the difficulty of accelerating and decelerating the actuators beyond the time lags caused by circuitry, mechanical linkages, and signal transit time. Inertia can cause over-shooting and oscillations about a target position. |
| Sponginess | A characteristic of pneumatic teleoperators in which controls and actuators are connected by a compressible fluid. To some extent, good controls can eliminate sponginess (see *stiffness*). |
| Backlash | The amount a control must be moved in the reverse direction before the commanded joint responds. |
| Stability | The ability of a teleoperator to move smoothly from one configuration to another and maintain it without jitter, hunting, or divergent oscillations. |
| Sensitivity | A teleoperator is sensitive if a slight motion of the controls causes actuator motion. Often "play" or a "deadband" will be built into the control subsystem to prevent excessive sensitivity. |
| Cross coupling | This occurs when commanded motion in one degree of freedom creates motion in another. The control subsystem design should preclude cross coupling. |
| Drift | Drift occurs when electrically and hydraulically actuated teleoperators may move slightly in a continuous fashion due to servo "leakage." |
| Compliance | The match between the manipulatory requirement of a task and the motion capabilities of the teleoperator (Fig. 5.1). Good control design can improve the dynamic match. |

| | |
|---|---|
| Reliability | The probability that the system will operate at some stipulated level of performance for a stipulated length of time. The control subsystem must help the overall teleoperator system meet reliability goals. |
| Fail-safe capability | When a teleoperator fails or loses power, the control subsystem should assure that the actuators retain their configurations. (Collapse could be disastrous in a man-amplifier.) |
| Self-protectivity | Limits switches and other control devices should prevent a teleoperator from damaging itself. |
| Cost | Self-explanatory |
| Power requirement | Power is critical in space and undersea work. The control subsystem should draw as little power as possible. |
| Support-equipment requirements | The total of all auxiliary equipment; such as repair and maintenance facilities, fuel-supply facilities and vehicles; and, of course, the trained technicians associated with this equipment. |
| Operator skill required | The effective matching of the man-machine interface can reduce skill requirements. |

## CONTROL THEORY

### Open-Loop Control

Imagine driving an automobile with the windshield blacked out and with no "feel" in the steering wheel. Without visual and force feedback, catastrophe would soon result. Control under these conditions is termed "open-loop," and though it would seem a disastrous approach to teleoperator control it is employed in special circumstances.

One such circumstance occurs whenever the control of a teleoperator is relinquished by the human operator to a preprogrammed set of instructions—say, a preprogrammed subroutine in an on-board computer that automatically stows a manipulator on a submersible. Open-loop subroutines are essential in supervisory control; in fact, the use of computers to relieve the operator in routine situations and provide special nonanthropomorphic skills is so important that we devote the next section to this subject.

Meanwhile, teleoperators that are normally operated in a closed-loop mode may revert to open-loop control under the following conditions:

1. If feedback is temporarily cut off, based on cues acquired before the displays were blacked out, an operator can usually make several movements safely. The feedback-deprived automobile driver mentioned above can, for example, pull safely off the road if he knows where he was before the blackout and if the traffic is light.

2. If feedback information suddenly becomes unintelligible due to noise or becomes too complex for the operator to cope with, the operator might well proceed open-loop fashion to some safe holding position.

3. If there is significant time delay and the operator cannot discern the consequences of his actions for several seconds, he may adopt a move-and-wait strategy in which each short open-loop move is prefaced by an analysis of the consequences of his last move. The operation of the Surveyor lunar surface sampler employed this philosophy. (See later section in this chapter on time delay.)

4. If a tedious repetitive task is anticipated, one cycle of the operation can be carried out once under closed-loop conditions, with all control information being recorded, and thereafter accomplished by supervisory control without the operator in the loop.

**Preprogrammed Control**

In preprogrammed control, the operator turns control of the teleoperator over to a machine, one with a memory that contains instructions for carrying out a given order. The instructions may be stored in a computer's memory or engraved in analog form on a grooved rotating disk or cam, like the famous Jacquet-Droz automatons in the late 1700s. The operator may transfer control by simply pressing a button, typewriter keys, or by reading a deck of punched cards into a computer. Or, in principle, the machine portion of the teleoperator may intentionally bypass the operator in an emergency and switch in a preprogrammed subroutine. A common feature of preprogrammed control is the absence of any feedback to an operator that would permit any modification of the action—the "manipulator stow" subroutine, for example. Once the subroutine is in action, it is played out. In other cases, the human operator can inhibit action and correct errors.

It is often desirable to initiate a subroutine which requires internal feedback of some sort (unseen by the operator) to carry out an instruction. An operator may in fact cut himself out of the loop and switch in a variety of supervisory subroutines, including: (1) the type of open-loop preprogrammed subroutine just described; (2) an automatically controlled closed-loop subroutine that utilizes feedback signals to reduce the task error, for example, the automatic movement of the teleoperator arms into configuration A; or (3) an adaptive or artificially intelligent subroutine that makes its own decisions on how to best carry out an operator's directive, perhaps by transferring object X to point B around an obstacle. Closed-loop subroutines (2) and (3) of course require feedback, whereas open-loop subroutine (1) moves ahead oblivious to feedback. In effect, we have established the matrix of operator-machine control relationships illustrated in Fig. 5.3.

|  | TYPE OF CONTROL | |
|---|---|---|
|  | OPEN-LOOP | CLOSED-LOOP |
| OPERATOR IN LOOP | MOVE-AND-WAIT STRATEGY | NORMAL TELEOPERATOR OPERATION |
| OPERATOR OUT OF LOOP | PREPROGRAMMED, SUPERVISORY CONTROL | ADAPTIVE CONTROL; ARTIFICIAL INTELLIGENCE; AUTOMATIC CONTROL; SUPERVISORY CONTROL |

**Figure 5.3** Matrix of various teleoperator control situations. To qualify as a teleoperator, the machine should operate with the operator out of the loop only in special situations where the human operator cannot cope with the task or where he wishes to relieve the task burden.

Is there a formal theory of open-loop teleoperator control? There is little to report here. Naturally, a strategy is important in an open-loop move-and-wait situation like that encountered in operating the Surveyor surface sampler. If one wishes to dig a trench on the Moon, one does not at first take big bites from an unknown medium that might damage the sampler itself. Instead, one devises a strategy composed of moves such as: Extending a sampler arm in increments of a half-inch at a time, waiting between moves to see the results.* In other words, adopt a "move gingerly" strategy.

Beginning with Ernst's work in 1961 (Ernst, 1961) several researchers, notably at M.I.T. and Case Western Reserve, have interposed a digital computer between the human operator and the manipulator (Verplank,

---
* Preprogrammed tapes controlled some Surveyor sampler arm motions.

1967; Beckett, 1967). The software and hardware employed in these NASA-supported experiments will be described later. A typical open-loop computer instruction during a stow subroutine or reactor core disassembly might be: Move joint C 5° clockwise. The computer merely acts as a switch in this case, turning on the motor driving joint C for the requisite number of revolutions. In open-loop control there is no feedback to assure the computer that joint C really rotated 5°, although a limit switch would probably be installed to indicate completion of the task.

### Closed-Loop Control

Sophisticated control systems depend upon feedback; teleoperator controls are no exception. Teleoperators are normally operated with man in the loop and with visual feedback. Even many of the supervisory subroutines that relieve man of participation in control depend upon internal feedback signals to carry out their instructions.

A large body of theory has grown up around the concept of feedback control (Savant, 1964; Gruenberg, 1967). Our objective here is to summarize some of the conventions and the general teleoperator approach.

The essence of feedback control is, of course, feeding some of the output back into the input to modify it. One tries to reduce the error with feedback, but sometimes this tactic is not successful and instability occurs. Some important control conventions are illustrated in Figs. 5.4 through 5.6. The first of the "block diagrams," Fig. 5.4 illustrates how an input, $R$, is affected by a control system element symbolized by the block and is algebraically represented by the transfer function, $G$. The $G$ symbolizes "something done" to the input signal. The output, $C$, is given by $C = GR$. The block diagram of Fig. 5.4 is completely equivalent to the equation $C = GR$. The control element thus represented is obviously linear. If two control elements are in series (Fig. 5.5), the

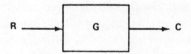

Figure 5.4  A simple open-loop control situation.

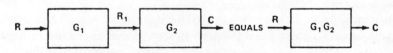

Figure 5.5  Two control components in series.

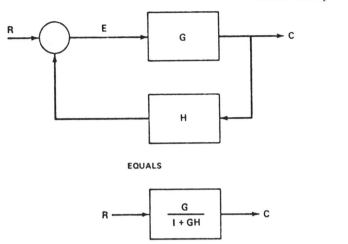

**Figure 5.6** A fundamental equality in control circuit theory.

overall equation is $C = G_1 G_2 R$. The $G$s are often called forward transfer functions and may include the human transfer function.

When feedback exists, control theory convention calls for the addition of the subtractor symbol, the circle in Fig. 5.6. Here, the input $R$ and the feedback signal, $HC$, are subtracted: $E = R-HC$; where $H$ is the feedback transfer function and $E$ is the actuating signal. Because $C = EG$, we can also write:

$$C = [\frac{G}{1 + GH}]R$$

This equation and Fig. 5.6 represent closed-loop control with negative (degenerative) feedback. Feedback can be positive as well as negative. Also, its frequency, phase, and other characteristics can be modified to achieve the goals of the control system designer.

For all the simplicity of Fig. 5.6 and the equivalent equation, they are really a facade for more complex equations describing the dynamics of the control system as measured in terms of its input and output voltages, displacements, or whatever the physical parameters may be. To transform the usually complex equation expressed in physical parameters into the G, H, R, C, E representation, one utilizes the well-known Laplace transform. Again, the reader should consult the many textbooks on control theory.

Conventional feedback control theory is applicable in principle to teleoperators with man in loop and when the teleoperator is controlled by automatic control subroutines that depend upon feedback, almost all extant teleoperators fall into these two categories. We specify "conventional" control theory because later in this chapter we will describe some new theoretical developments oriented specifically toward teleoperators.

### Adaptive Control and Artificial Intelligence

The word "adaptive" is employed fairly loosely in the control literature. Generally, an adaptive control system is one which adjusts to meet changing circumstances. In this sense, any feedback control system is really adaptive. In this book, however, we narrow the meaning to include only control systems that can cope with changing external circumstances beyond the capacity of simple feedback control. Two examples: avoiding an obstacle and finding the quickest way to take a manipulator from configuration state A to state B. In other words, judgment and decision-making are involved in adaptive control; something beyond the ken of a "deterministic" feedback control system such as a thermostatic temperature regulator. The distinction, however, is rather fuzzy.

Even fuzzier is the distinction between adaptive control and artificial intelligence. An artificially intelligent machine would not only be adaptive but would also have the ability to learn from past mistakes and be able to devise strategies of a general nature to reach goals set by itself —or perhaps goals set by man if the machine still depends on him at this stage of development.

A teleoperator, being a man-machine system, is always adaptive and intelligent when man is in the loop because man has defined these characteristics from his analysis of himself. But when operating in a subroutine, we often look to the machine portion of the teleoperator to do a little thinking for itself.

To illustrate how feedback theory also applies to subroutines, we describe how the computer-controlled manipulator at Case Western Reserve assures that it has correctly carried out an instruction (Beckett, 1967). If the subroutine requires that the manipulator move to a specified configuration (or "state"), the computer compares the current configuration of the manipulator, axis by axis, with the desired configuration. The differences in axis positions are converted into analog voltages. These voltages—really error signals—drive the axis motors until the errors disappear. The resultant configuration should be the desired one since all errors have been nulled. The feedback in this example consists of the voltages (from axis potentiometers) representing the manipulator configuration as a function of time. It is classic feedback control. The operator, though, is not in the loop during this operation.

The Case computer-controlled manipulator also exhibits a kind of adaptive behavior in its ability to avoid obstacles in its path. If the computer memory knows the location and configuration of the obstacle situated between the initial and final manipulator configuration, it will first check to see if other terminal arm-hand configurations can place the hand in the right position. If so, the obstacle may be avoided by proceeding to one of these directly. The computer checks to see. If obstacle avoidance is still impossible, the computer will explore several paths leading around the obstacle, select the one requiring the least transit time, and set the manipulator in motion along this path. Clearly, a judgment and a decision have been made.

Similar obstacle-avoidance studies are being pursued under NASA contract by Sheridan's group at M.I.T. using sets of heuristics arranged according to a priority criterion (Sheridan, 1967). One heuristic approach might be to try a series of straight line motions tangent to the obstacle's peripheries.

Most of the walking machines we see today are preprogrammed and open-loop. They tread away blindly, regardless of the terrain. R.J. Hoch and his associates at Battelle-Northwest Laboratories have conceived of a method that may make walking machines adaptable to varying terrain (Hoch, 1967). The germ of the Battelle idea lies in the short-term memory of a small computer and the quantizing of the control system. Control of the walking machine by pistons is accomplished by a series of discrete pulses, N pulses per second to each piston. Initially, the control pulses would be those that would carry the vehicle over ideal nonvarying terrain at the gait and speed set by the operator. In this mode, the operation would be preprogrammed; but as the terrain departs from ideality, the piston backpressures (the discrete feedback pulses) would also depart from those expected from an ideal terrain. The differences between the ideal and the real signals would be stored in the control computer memory and used to modify subsequent control pulses. The older the differences the less their weights in determining the next cycle of control pulses. The Battelle scheme would also employ sensors that feed back data on vehicle stability that may bypass the normal controls in favor of some emergency subroutine—say, one that prevents the vehicle from overturning. The use of past deviations from ideality in determining future action is a form of learning. We humans are adaptive walking machines, except that we can usually see the terrain ahead and add this knowledge to that from past experience. Note that the Battelle walking machine would not have the human operator in the loop while controlled by the computer, it would operate under supervisory control during these periods.

### The Time Delay Problem

Many people have experienced the disconcerting effects of delayed audio feedback, particularly in public address systems. Delayed visual and force feedback can compromise teleoperator control in a similar fashion. Here, we define the problem and look at some "preview control" models; later we will cover the predictor displays that have been designed to help solve the time delay problem.

NASA has been concerned with transmission delays resulting from the finite speed of radio signals over the great distances in outer space. Between Earth and Moon, the round-trip signal time is roughly 2.6 seconds—enough to disconcert an Earth-based operator of a lunar machine (Fig. 5.7).

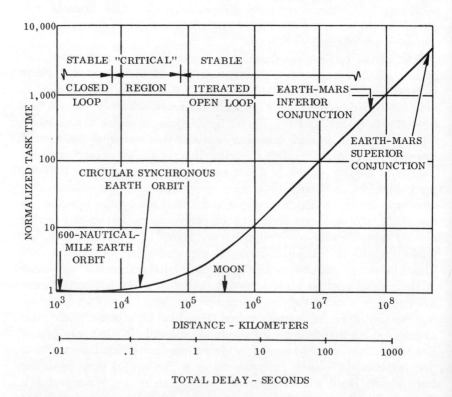

**Figure 5.7** Normalized task time versus total time delay. A critical region, around 0.25 sec, where operator confusion is possible, occurs in space missions at very high orbits. Move-and-wait operator strategy would be a successful but slow strategy for work on the Moon and planets.

Besides the signal propagation time delay, feedback information also encounters electrical circuit and mechanical device delays. The operator in the control loop also slows signals down. Wargo has summarized human delays for one-choice situations as follows (Wargo, 1967):

| | |
|---|---|
| Receptor delays | 1–38 milliseconds |
| Afferent transmission delays | 2–100 |
| Central process delays | 70–100 |
| Efferent transmission delays | 10–20 |
| Muscle latency and activation delays | 30–70 |
| Total delay | 113–328 milliseconds |

These figures are for alert, pre-warned subjects. Reaction times lengthen when the unexpected occurs and when the choice is a complicated one. Human delays and equipment delays are usually much smaller than those NASA anticipates from propagation delays in space exploration.

The portion of the overall time delay that disconcerts the operator is that part that prevents him from seeing the immediate consequences of his actions. Ferrell and others have summarized past work in the field of delayed sensory feedback (Ferrell, 1965; Arnold, 1963; Leslie, 1966). The early studies involved tracking experiments of various kinds with delayed visual feedback. All of the studies concurred that time delay was deleterious to performance. In connection with its projected remotely controlled lunar vehicles, NASA has sponsored work at Stanford University (Adams, 1962) which indicated that driving performance worsened with increasing time delay. The situation deteriorated faster as vehicle speed increased, as the vehicle course became more complex, and as the television field of view narrowed. Similar effects have been noted for delays in auditory feedback. Recent theoretical work by W. H. Thompson at ANL indicates that force feedback, too, is of diminishing utility as time delay increases (Thompson, 1968).

Also applicable to teleoperator control were Ferrell's 1963–1964 experiments with a two-dimensional manipulator with variable time delay in the visual feedback. The manipulator operators in this case were able to pace their activities and work out strategies that suited them best. The tracking and vehicular experiments, in contrast, force the operator to synchronize his activities with the input signal. Ferrell found that task completion time increased with both time delay and task difficulty. Some important conclusions from Ferrell's work were:

1. Both operators independently adapted a move-and-wait strategy as the best and least confusing solution to the unnatural time-delay situation.
2. There were no unstable or oscillatory movements. This fact was attributed to the adoption of the open-loop move-and-wait strategy.
3. The operators found the work tiring and difficult, but not emotionally upsetting as other operators have reported for forced-pace time-delay tracking experiments. Trials where these operators were asked to use a move-slowly strategy did, however, disconcert them.

In summing up, time delay (and task difficulty) can be overcome by taking additional time—mostly waiting time between successive operator moves and the returned feedback.

We shall see later how the augmentation of the human controller by a computer in supervisory control can help to overcome instabilities induced by time delay.

### Manual Control and Tracking Theory

Earlier in this chapter, we have occasionally mentioned tracking theory —perhaps a little too disparagingly. Nevertheless, modern manual control theory is largely built upon a foundation of tracking experiments. These quantitative experiments form the only real basis for evolving and testing hypotheses in manual control. And manual control theory is the only kind of control theory we have that includes in the loop the human operator with all his idiosyncracies.

Three main types of tracking are recognized:

1. Pursuit tracking, wherein the operator sees both the moving target and his own corrective responses. A common analogy is a duck hunter using a gun with an open sight. In laboratory practice, the operator tries to follow a moving target, say, a moving spot, using a joystick or some other control.
2. Compensatory tracking, in which the operator sees only the *differences* between the moving target and his response; i.e., the error. In this type of tracking, the operator attempts to null the difference signal.
3. Precognitive tracking, which exists when the operator has complete information about the target's future—as in shooting at a duck in a shooting gallery. In the true sense of the word, this is not really tracking.

Which of these kinds of tracking have application to teleoperator control? Pursuit tracking applies if the teleoperator is trying to pick up or perhaps hit a moving object, a very rare situation in present-day teleoperator practice. In manipulation, the targets are generally stationary; so is the environment. In picking up an object, the operator first directs the manipulator hand to the general area of the target in a gross movement; then, in a series of fine adjustments, the hand is accurately positioned and the jaws closed. The same kind of coarse-fine "tracking" occurs when the target is moved from position A to position B. But which of the three main varieties of tracking describe the situation best? Obviously, precognitive tracking is closest, but it is not bona fide tracking at all. No formal theory exists for precognitive tracking.

There are, however, elements of pursuit tracking that may be applicable. For example, the first gross movement of the manipulator hand to the region of the target is akin to getting the duck in the gun sights, and the fine motions prior to grasping the target really involve nulling out the position errors the operator sees visually. In sum, there is no single type of tracking that seems to cover teleoperator action. Further, there is no theory at all that really grasps the planning and strategic thinking of the person controlling a teleoperator.

These things being so, why bother to discuss tracking theory at all? The answer must be that tracking theory gives us the only quantitative insight into the behavior of humans in control loops, despite its acknowledged drawbacks. Any comprehensive theory of teleoperator control (including the operator) must build on (or alongside) manual control theory.

## Some Approaches to Teleoperator Control Theory

Granted the weaknesses and general inapplicability of classical manual control theory to teleoperators, what has been done in the way of formulating a useful description of teleoperator control processes? Not a great deal! This should not be too surprising, because the human functions of planning and strategy setting are still being argued and have yet to be embraced by mathematics.

Seidenstein and Berbert (Seidenstein, 1966) have examined the extant literature in those areas which they believe comprise the most important "extra" dimensions of teleoperator control:

1. Judging the best path for approaching the target.
2. Approaching the target and minimizing undershoot and overshoot.
3. Orientation of hand for manipulatory task.
4. Final adjustment of arm and hand.

In 1966, Seidenstein and Berbert found essentially no important literature that would give a foundation upon which to build a comprehensive theory of teleoperators. However, their literature review did not encompass K. U. Smith's work and much of the psychomotor theory of perceptual and motor organization (see Bibliography). Some of these theories may ultimately prove extremely useful in teleoperator theory.

Significant direct attacks on the teleoperator problem have been made by Sheridan's group at M.I.T. and Lyman's group at UCLA (Sheridan, 1967; Lyman, 1967). Both groups bypass the human operator as a planner and strategy formulator. Although man still retains "executive" control of the operation in their approaches, their theories concern only the machine part of the teleoperator in supervisory control situations.

One of Sheridan's students, D. E. Whitney, has completed some pioneering work in the field of supervisory manipulation in state space (Whitney, 1968). He lists the following attributes of a good computer-controlled manipulator, which sound remarkably like the qualities a human manipulator operator must have:

1. It employs a symbolic representation or model of the task site. All objects, obstacles, fixed support surfaces and effectors (jaws, tools, etc.) are represented in their proper spatial relationships.

2. It can identify goals in this model. A goal may be thought of as a particular configuration of the objects, obstacles and effectors which the operator wishes to attain.

3. It understands how the effectors can alter the task site as well as how these alterations are represented in the model.

4. It can receive commands which specify goals to be achieved and constraints to be obeyed. Then, using items 1, 2, and 3, it can translate the command into an expanded equivalent. ("Expanded" means that strings of manipulator primitive commands have been substituted for the human primitive command; "equivalent" means that these manipulator primitive commands, when carried out, can be expected to accomplish the previously stated goal.) In other words, the system can make a plan for carrying out the task.

5. It can execute this plan, judging its progress against the plan's expectations, keeping track of its progress by updating the model, and asking for help if trouble develops or things do not go according to the plan.

Suppose the manipulatory task is to move an object from one point to

$$x(t_0) = \begin{bmatrix} x_0 \\ y_0 \\ 0 \\ 0 \end{bmatrix} \quad \text{to} \quad x(t_f) = \begin{bmatrix} x_f \\ y_f \\ 0 \\ 0 \end{bmatrix}$$

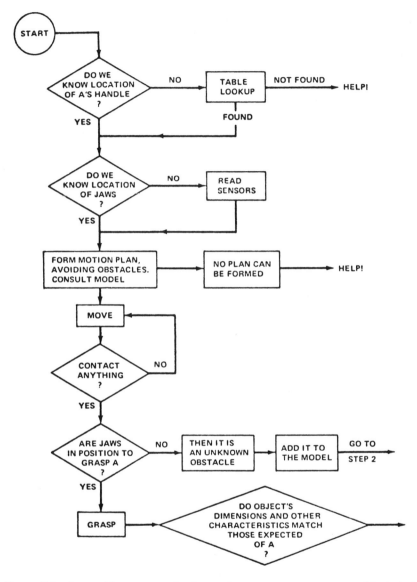

Figure 5.8  A possible program for instructing a computer to carry object $A$ to location $X$.

another on a table, avoiding an obstacle on the way. In state space coordinates this means moving from between time $t_o$ and time $t_f$. The zeroes in the state vectors represent velocity components. The values of $x$ and $y$, however, must not assume values off the table or too near the obstacle. The best trajectory is found by testing each trajectory between the two points which satisfy the constraints and comparing one against the other using some time, distance, or cost criterion. Problems such as this are common in engineering and are solved by the methods of "optimal control," including the calculus of variations, dynamic programming, etc.

A typical manipulatory task—carrying object $A$ to location $X$—can be programmed as shown in Fig. 5.8. Note how the computer must be instructed in details that men understand instinctively.

"Inhibitory control," proposed by Lyman and Freedy, assumes that some paths between initial and final states (teleoperator plus environment) are more likely than others. Our eating motions, for example, are rather stilted, and this is true of most routine manipulatory functions, especially those by wearers of artificial limbs. In inhibitory control, an adaptive controller—probably a computer—would drive the teleoperator between the initial and final states (initially selected by the human controller) along historically favored paths. The human operator monitors the motion and inhibits it where it is in error, due perhaps to a new obstacle placed in the environment. The adaptive controller adds this new information to its running account of favored teleoperator motions; while the human operator still monitors the activity, he is relieved of the burden of planning and detailed execution of the task. This approach differs from that of the M.I.T. group in that human judgment guides the choice of path rather than some optimal control scheme based on minimum time or some other constraint.

### Application of Control Theory to Unilateral Teleoperators

Unilateral teleoperators (sometimes called "rectilinear" in error) are controlled by the following:

1. Switches or potentiometers which actuate motors driving the various degrees of freedom. Feedback in this instance is visual as the operator corrects errors in position and orientation. If switches are used, this control technique is called "rate" or "velocity" control; potentiometers permit variable motor speeds or "proportional rate" control.

2. Replica or prosthetic-type controls that are analogs of the actuator subsystems. The servomotors driving the various degrees of freedom are actuated by an error signal which is proportional to the difference be-

tween the desired configuration specified by the controls and the actual configuration of the teleoperator. This is termed "position control."

In any real teleoperator control system, the differential equations describing the motion of the arms, hands, and legs are complicated by the fact that these appendages have mass (which may lead to overshooting the target), have friction in the joints, and may move in an appreciably viscous medium, such as seawater. Such considerations are part and parcel of the design of most control systems, such as those of radar antennas and guns. Thus, the system may be damped to reduce overshooting or "hunting," yet excessive damping will cause undesirably sluggish response. A compromise must be found. The reader should consult texts on control system design. Only a few specialized reports have been published on the application of control theory to specific teleoperators (Whitney, 1968; Pieper, 1968).

The just mentioned theory, though well-developed, excludes the most important control-loop component: the human operator. Since we have no practical, analytical way of incorporating the human operator into the teleoperator control equations conventional control theory remains only a guide. All is not lost, though; because when the mathematics become intractible, insight can frequently be gained by simulating control systems electromechanically. With a human operator plus a reasonable analog of the electrical and mechanical components, different control schemes can be compared, stability regimes can be investigated, and even the analytically elusive properties of the operator can be studied.

Unfortunately, little basic simulator work has been completed. The most significant studies are those by Ritchie, Inc., under Air Force contracts (Seidenstein, 1966; Williams, 1966), and at General Electric, under DOD sponsorship (General Electric, 1968). The Ritchie simulators employed three and four degrees of freedom and incorporated manipulator arm mass, damping factors, and motor characteristics. The following conclusions are taken from the Ritchie studies:

1. Proportional rate control is better than fixed rate control in terms of task time and efficiency.
2. The use of high speeds in approaching the target results in longer fine adjustment times.
3. Small targets require higher travel times than large targets with fixed rate control, but the opposite was found with position control—an "illogical" result.
4. There seems to be an optimum rate of motion (4 in/sec) for the conditions of the experiment. Overall performance decreased above and below this rate.

## Application of Control Theory to Bilateral Teleoperators

The first bilateral servoed teleoperators were built by R. C. Goertz's group at Argonne National Laboratory (ANL) in the early 1950s. Several key theoretical papers originated from this work (Goertz, 1953; Arzebaecher, 1960) which was very extensive. We can show only the general approach here. Following Burnett (Burnett, 1957), the somewhat idealized symbolic diagram for the ANL Model 2 force-reflecting electrical

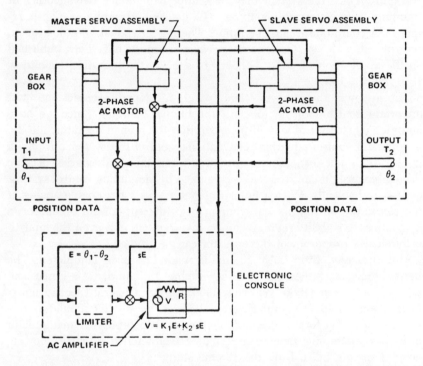

Figure 5.9 Symbolic diagram of the force-reflecting servo used in the ANL Model 2 electric bilateral master-slave.

master-slave manipulator is presented in Fig. 5.9. The dynamic equations in Laplace transform notations are:

$$T_1 - K_T I = (Js + F)s\theta_1$$
$$T_2 + K_T I = (Js + F)s\theta_2$$
$$V = K_1 (\theta_1 - \theta_2) + K_2 s(\theta_1 - \theta_2)$$
$$IR = V + K_b s\theta_1 - K_b s\theta_2$$

where  $J$ = the motor and gear train inertia
 $F$ = the mechanical viscous friction
 $s$ = the Laplace transform variable
 $K_T$ = the torque constant
 $K_b$ = the back EMF constant
 $T_1, T_2$ = externally applied torques
 $\theta_1, \theta_2$ = angular displacements
 $R$ = resistance
 $I$ = current
 $V$ = voltage
 $K_1, K_2$ = constants defined in Fig. 5.9

The first two equations are torque equations; they assume linearity and complete bilateral symmetry. The equations represent what is termed a first-order (linear) analysis. Stability is indicated, but higher order (non-linear) analysis could reveal instabilities leading to oscillations.

When bilateral joints are connected in series the analysis gets even more complex. The arms of Hardiman I have three such joints. The signal flow and block diagrams are too involved to reproduce here, and the reader is referred to the original General Electric report (General Electric, 1968). In fact, General Electric did not try to analyze the three-joint bilateral model; instead the engineers extrapolated the results of the three-joint unilateral and single joint bilateral cases. The three-joint bilateral model was simulated on an analog computer. It was found that the compensation networks described earlier for the unilateral case also stabilized the bilateral model from 0–1500 pound loads. Teleoperators like Hardiman I are feasible according to the General Electric study; however, during the design process, provisions should be made for adjusting the proportional, rate, lag, and velocity feedback terms over wide ranges.

Apparently, any practical, real-world teleoperator will defy rigorous analysis by virtue of its complexity, at least until better analytical techniques are worked out. The presence of a non-linear, time-varying human operator only worsens the prospects. Thus, the major conclusion of this chapter must be that pure analysis can only guide teleoperator design in terms of pinpointing design weaknesses and helping the designer think out and grasp the interrelations among control parameters.

The situation is not hopeless because even the three-joint bilateral case can be simulated. Even better is an engineering mockup of the teleoperator with a human at the controls. A good, general approach to teleoperator control design would be threefold: (1) limited analysis; (2) simulation, and (3) engineering mockup.

## THE MAN-MACHINE INTERFACE

### The Nature of the Problem

As computers and other machines assume more and more importance in our lives, the body of literature discussing the *man-machine partnership* and *man-machine symbiosis* grows. In the preceding chapter, it was obvious that the control-theory describing the total man-machine system is rudimentary at best. We try to describe man with the same kinds of equations we use for machines, but success still eludes us. In teleoperator theory, man and machine seem analytically irreconcilable; and to make the schism seem more complete few men doubt that they are superior to machines in many important ways. Yet, man and machine must be integrated, especially in the teleoperator where the partnership is closer than it is in most man-machine systems.

Man and machine meet at two hardware interfaces in the teleoperator: the controls and the displays. Specific control and display hardware are covered in the next two chapters. In this chapter, the general, more philosophical problems of matching man and machine at these two points will be discussed.

Should we match man to the machine or the machine to man? (This question is overworked in today's literature.) The answer, of course, is that we do both to that degree needed for best teleoperator performance. However, because we still do not understand machines well and know ourselves even less well, this brave plan cannot be consummated easily—and then only very imperfectly. Even in our ignorance, though, we can approach the problem in an orderly fashion by: (1) describing the pertinent properties of man and machine; (2) rationally allotting tasks to one or the other; and (3) building sound bridges across the interfaces at the controls and displays.

### Defining the Human Operator and the Machine

Whenever a subject is either controversial or not amenable to precise description, the literature is abundant; this is the case with man as a controller. Fortunately, a recent and thorough survey of this field has been completed by Serendipity Associates under a NASA contract (Price, 1968). We lean heavily on this survey, making use of those portions applicable to teleoperators.

We are trying to define the man-machine interface and just where man or machine should assume responsibility in a teleoperator. To this end, we list pertinent man and machine attributes side by side for the sake of easy comparison.

## Sensory Comparisons

| Man | Machine |
|---|---|
| Senses limited to narrow ranges. However, these limitations do not affect teleoperator control significantly (except in undersea work) because properly designed displays can overcome most limitations. | Sensory ranges extend far beyond those of man. A machine can also sense X-rays and other environmental factors normally invisible to man. |
| Man's input channel capacities in all senses are limited. They can be saturated easily. He may need machine help at times. | Machine channel capacities can be made as wide as desired at a price measured in power, weight, cost, etc. |
| Resistant to jamming and noise. Man can often filter out the signals he wishes to use. | Generally more subject to jamming and noise. |
| Man can sense and recognize patterns, color codings, and written or printed characters. Targets can often be discerned amid noise and clutter. | Pattern recognition possible, but not well-developed yet. |
| Man is usually considered to be a single-channel detector at any given instant, implying that he must switch his attention from one channel to the other. However, sight, sound and touch usually work together easily in manipulatory tasks. | Machines can handle many channels simultaneously. |
| Man's sensory capabilities are affected by fatigue, general health, noise, and other environmental factors. | Machines are less affected by the environment and wear. |
| Man's senses cannot be calibrated reliably in absolute terms to provide quantitative data. | Instruments can be accurately calibrated and easily read. This may be an advantage in delicate manipulations. |

## Sense Interpretation Comparison

| | |
|---|---|
| Humans often see only what they expect to see and can be fooled by such things as optical illusions. | Machines are much more literal in their interpretive functions. |
| If a new, unexpected situation (a new "universe") is encountered, man can cope with it better than a machine. An emergency or accident would fall in this category. | Generally, machines can deal only with the known and expected—the known "universe." |
| Man's interpretation of data depends upon his previous history with them. Experience is usually beneficial, though it can prejudice an operator. | Historical information can affect interpretation by machine only in those ways which can be implemented by computers; i.e., time averaging, etc. |
| Man's reliability as an interpreter depends upon his emotional state and fatigue. | Machines are more objective, tireless, and unemotional. |
| Written language, color codes, and other symbols are readily interpreted. This is particularly useful in handling coded objects. | Languages, codes, and abstract symbols can be interpreted only with difficulty. |
| Given the symptoms, a human can troubleshoot a malfunctioning teleoperator. | Machines can also do this but only to a limited extent. |
| The human operator can hypothesize. He can ideate. He can suggest alternative modes of action. | Machines cannot do these things well. |
| Men are poor monitors of infrequent events. | Machines are much more reliable as monitors. |
| The human operator is poor at monitoring continuous signals and processes over long periods of time. | Machines are so good at monitoring that some have suggested that they be employed to monitor men instead of vice versa. |
| Man is good at detecting deviations from normal, particularly in the presence of noise and other signals. | Machines are better than men at monitoring simple processes, but they are less successful when patterns and symbols are involved. |

## Information Processing Comparison

Relatively low-speed information processor. Essentially a single-channel processor at any instant.

High-speed information processor. Can handle many channels simultaneously.

Weak and inaccurate as a computer. Tires quickly; especially in routine, boring jobs.

Tireless and fantastically accurate in comparison to man. Man should never compute if he can get a machine to do it.

Man is easy to program. He does not require extremely precise instructions. He is *flexible*.

Programming machines is time consuming. Each instruction must be detailed and specific.

Information can be processed in a wide variety of formats. Special coding, punching, etc., not necessary.

Computers are very specific and limited in the forms of information they will accept.

Man's bandpass is about three radians per second. He can transmit 30-35 bits/sec.

A machine's bandpass and data rate can be made much larger than man's—at a price. A machine can thus potentially manipulate much faster than man.

Man's short-term memory is limited in size, accuracy and permanance. Access time is relatively high.

Machine memory can be almost unlimited. Accuracy and performance are high. Access time is very low.

Man processes information so slowly that he is relatively inefficient in search tasks, although he is good at recognizing and identifying targets once they are located.

Machines can rapidly search huge quantities of data for well-defined targets, but accuracy suffers as target definition is worsened.

Man has an excellent long-term memory for related events. Generalized relevant patterns of previous experience can be recalled to solve immediate problems.

This property can be built into machines only at great expense.

## Decision-Making Comparisons

| | |
|---|---|
| Man can generalize and employ inductive processes. | Machines have less capability for induction and generalization. |
| A human being does not always follow an optimal strategy—usually because he cannot perceive or examine all ramifications of a situation and cannot compute all the possible solutions. | Machines always follow built-in strategies, or they can compute optimal strategies given sufficient information. |
| Decisions can be made despite incomplete information and where the rules are not certain. | A computer usually demands complete information before making a decision. |
| Human decision-making time is relatively high. Often man wavers between alternatives if the decision is not clear-cut. | Machines are fast and specific. |
| Man is always needed to set priorities, establish values, set goals, risks. | Machines must be instructed as to priorities, values, goals, etc. |
| Targets of opportunity are recognized better by man. | Machines are relatively insensitive to unspecified opportunities. |
| Humans can improvise superbly. | Machines improvise poorly. |
| Man learns from past experience. | Machines can learn, too, but are not proficient at it yet. |
| Human operators prefer tasks with high degrees of responsibility and authority. Pride and a need to prove "human value" are factors here. | Degrees of responsibility and authority are irrelevant to machine. |

Given the attributes of man and machine, how does one draw the interface between them? In practice, this question is answered by allocating tasks or portions of tasks to each. The type of machine and the job to be done are important in determining how much man and how much machine will be employed in control. Not too surprisingly, the personal philosophy of the designer has something to do with establishing the interface: Some engineers believe computers should be brought in for supervisory and preview control, others want men in the loop at all times. Finally, no matter how carefully control tasks are apportioned between man and machine, the match will never be perfect.

## Controlling Comparisons

Cannot exert large well-controlled forces. (Force or pressure is man's primary control mechanism.)

Machines can exert considerable force with speed, steadiness, and precision. Reaction time is much smaller than man's.

Superb at manipulation, construction, creative work, non-routine tasks.

Good at routine and well-defined tasks; i.e., those performed under supervisory control.

Tires quickly. Easily bored by routine, repetitive tasks. Man is easy to overload.

Tireless, never bored, hard to overload.

Man's motor output seems to have a bandwidth of about 10 cycles per second, with a natural periodicity (to be avoided) of $\frac{1}{2}$ to 1 cycle per second.

A machine can be designed for almost any bandwidth if one is willing to pay the price.

The motions possible with the human body, though marvelously contrived, are limited in amplitude and articulateness—some motions are impossible, such as telescopic extension of limbs.

In principle, machines are not limited in amplitude and articulateness of motion.

Performs well in emergencies. Can take remedial measures. Man is adaptable and can "reprogram" himself.

Machines do not adapt well to emergencies. They either stop or plod blindly ahead.

Man is often nonlinear in his manipulation of controls.

Linearity or any other function can be built into machine controls.

Humans are highly variable in physique and capability (Fig. 31). Allowance must be made in interface design for this variability.

Machines can be built with fairly well standardized interfaces.

To paraphrase the Biblical quotation: Render unto the machine the things that are the machine's. In the very specific area of teleoperator control, the problem of task allocation is rather simple because today's teleoperator normally works with the human operator in *full real-time* control of *all* activity; that is the operator usually renders nothing to the machine in terms of control. Of course, much of the sensing and actuating is done by machine, but no direct control functions are carried out unless the operator takes himself out of the loop and institutes a subroutine. Excluded from this generalization are the many *local* control loops that exist in any sizeable machine.

When the first master-slaves were developed in the late 1940s, the human operator was essentially in full control of every operation—lifting, moving, pouring, manipulating. In fact, the word master-slave is rather contemptuous of the machine role in the man-machine partnership. Before long, however, drills, saws, hammers, and other tools were being used in the same way man uses them. In other words, the human operator began to depend upon the machine for laborious "subroutines." No one thought to call a slave-held drilling operation a subroutine, but nevertheless the operator did relinquish part of the control task to the machine. Tools have become more and more important in the effective application of teleoperators; and the most important of these tools is the general-purpose digital computer. The computer is somewhat like man in the way he thinks, but it is undeniably on the other side of the man-machine interface.

Few systematic objective rationales exist for drawing the man-machine interface. Obermayer and Muckler have classified past attempts into five categories or, more properly, philosophies (Obermayer, 1965):

1. Automate wherever the task can be described in sufficient detail for engineering design, even though man might do some of these tasks better. Under this philosophy, man is assigned poorly defined and complicated tasks. Result: poor use of both man and machine.

2. Follow traditional roles and preferences, wherein man serves as the prime controller of vehicle attitude and power (as on aircraft). This approach has generally been discredited as man-machine systems have become more powerful and complex.

3. Assume specific human capabilities and limitations and design to make the best use of man under these conditions. Usually, this has been interpreted to mean that man should be used only as a narrow-band, simple amplifier. In teleoperator work, man is obviously much more than this, being a strategist and decision-maker as well as a supplier of control forces.

4. Assume a formal mathematical model of man (a human transfer function) and design the control system as if man were a completely specified servo element. This approach is hardly applicable to teleoperator design.

5. Make a direct empirical assault on the systems with simulators. Because some teleoperators are so complex; viz., Hardiman; this is occasionally done.

None of these five approaches really suffices for teleoperator design; they do not take into account man's most essential attributes—planning, decision-making, etc.—all difficult to reduce to equations.

In the context of future teleoperator design, the man-machine interface seems to be moving in the direction of letting the machine do more of the work. The following general assertions emphasize this trend and also offer some general guidelines relative to establishing the man-machine interface in teleoperators.

**Assign to Man These Control-Oriented Tasks:**
Pattern recognition
Target identification
New, exploratory manipulation
Long-term memory
Trouble-shooting, emergency operation
Hypothesizing, ideation, planning
Interpreting variable format data
Inductive thinking
Setting goals, priorities, evaluating results

**Assign to Machine These Control-Oriented Tasks:**
Monitoring multichannel input
Boring, repetitious manipulation
Precision motions and precision force applications
High-speed motions, particularly oscillatory
Short-term memory
Computing
Monitoring
Deductive analysis
Development of optimal strategies
Nonanthropomorphic motions

The tendency today is unquestionably to let the machine portion of the teleoperator do the hard, repetitious work, while the human thinks, plans, and explores. As machines become better able to identify and manipulate targets in accordance with man's general instructions, the machine will take over even larger portions of the control task.

The philosophy of applying machine (computer) control wherever reasonable has a profound effect upon the design of displays and control hardware. If man is to adopt more and more the function of an executive, he will need more executive-type controls; that is, controls that switch in subroutines. A specific subroutine could be initiated by a switch, a coded signal, or even voice command. Supervisory controls, therefore, are abstract and far-removed in terms of spatial correspondence from the master arm of a master-slave manipulator. The more "intelligent" the machine, the more abstract the controls and the less often man would enter the control loop to operate the controls directly. At the far end of the spectrum—where the true robots dwell—today's crude anthropomorphic master arms and hands would be replaced by general verbal instructions.

## BRIDGING THE INTERFACE

Once control tasks have been divided between operator and machine, there remains the "communication problem," which means insuring that man can command the machine efficiently and that the machine can feed back information to man with ease. The two points of contact where matching is necessary are at the displays (machine output) and the controls themselves (man's output).

For effective control of the teleoperator, many engineers believe that the controls should be organized like man; that is, be anthropomorphic—a true extension of man. This interface-bridging philosophy is different from, if not opposed to, the school that wants to make fuller use of machine capabilities and supplants anthropomorphic controls with switches that initiate machine-controlled subroutines. Obviously, the more the machine is in command, the less anthropomorphic the controls need to be. In hardware, these two philosophies are represented by the ANL electric master-slaves with slaved TV display on one hand and the largely computer-controlled manipulators at Case Western Reserve and M.I.T. on the other. In between are a few manipulators displaying various combinations of anthropomorphism and the more abstract, symbolic controls. This bifurcation of the field is becoming more evident each day.

There are undeniable advantages in anthropomorphism and spatial correspondence,\* the two prime tenents of the make-the-machine-like-man school:

1. An operator can use skills learned in everyday life to run the machine.

2. Operation is natural, not abstract, and requires less training.

3. Tasks requiring a high degree of physical coordination are often possible; e.g., hula-hooping.

---

\* Spatial correspondence exists when a motion by the human controller is duplicated by the machine.

4. The teleoperator is "generalized," like the human operator, and is readily adaptable to many varied tasks.
5. The operator "feels at home." He identifies himself better with the task.
6. It is suspected but not objectively proven that an operator's reactions in emergencies are quicker and more effective.

The computer-oriented school employs man's higher powers—planning, decision-making, etc:—and matches these abstract outputs to the machine using codes understood by the machine. Using today's technical vernacular, the first philosophy matches hardware to the human; the second matches software to the human. Some advantages and disadvantages of an abstract, software dialog are:

1. Abstract language can communicate more control information per unit time to the machine.
2. Nonanthropomorphic commands, such as wrist rotation, can be given.
3. Man is not "wasted" in dull, routine tasks and can use his higher faculties to do a better job.
4. The abstract language is usually highly specailized and may not meet the requirements of the task, especially an emergency.
5. Repetitive tasks can be done with high speed.

We have discussed so far matching the machine to the human operator; perhaps the operator is malleable too. Operator selection and training can help bridge the man-machine interface. Operators should be selected with the same care as for jet pilots. Factors such as depth perception, eye-hand coordination, and reaction time are important in the operation of contemporary master-slaves. Good physical condition is also a prime requisite because remote manipulation is arduous, demanding work. Training with manipulators or simulated tasks is essential. Although a few minutes with a master-slave can give a novice a good feel for the machine, only many months of experience will make a good operator. At the other end of the teleoperator spectrum, one would suppose that operating a teleoperator possessing a large array of subroutines would require high analytical power and abstract reasoning capabilities. But none of these man-and-computer-controlled teleoperators has been used operationally as yet.

Summarizing these points, we note that the man-machine interface is at best a poorly drawn boundary between man and machine, particularly in teleoperators. It is a dynamic boundary that changes with the application, with the machine state of the art, and even with the personal philosophy of the teleoperator designer. There is no detailed, well-defined, objective rationale that tells a designer how to deal with the man-machine interface. There are rules-of-thumb, opinions, and checklists for thinking out the problem. Basically, the field of teleoperator design is too young for hard and fast rules. Despite the lack of rigor evident above, the man-machine interfaces has been surmounted many times in many ways during the past two decades. In the next two sections, we relate some of the solutions—past, present, and future—in terms of control and display hardware.

## TELEOPERATOR CONTROLS

### An Overview

The fundamental function of switches, joysticks, and other control hardware is to translate the commands of the operator into signals that can be understood by the machine portion of the teleoperator. In control engineer's language, a control is really a transducer—a device that converts a signal from one form to another; for example, the force on a joystick to a proportional voltage operating a motor. The signals generated by the control hardware may by simply proportional to some physical input from the operator, or they may be symbolic; that is, they may contain coded meanings, such as move from point $A$ to point $B$. A simple symbolic input, perhaps generated by a typewriter, can release a subroutine containing a long train of "primitive," low-level signals to the teleoperator's basic actuators.

Man's signals to his machines are usually generated by his hands. Direct force activates most teleoperator controls, including those that switch in subroutines via a teletypewriter. Force and pressure from man's appendages also configure complex controls, such as analog or replica controls, or activate arrays of switches in complex patterns. In some cases, particularly in the medical field, control forces are created by the feet, the tongue, the head, and various muscles throughout the body. Man also generates more subtle outputs: eye movements, muscle bulges, and electromyographic signals from electrodes attached to the body are used for control purposes. Finally, the human voice can carry a heavy traffic of control information if we can find a machine that can listen and interpret properly.

Regardless of how the human body creates control signals, they can be classified into four types:

1. On-off signals, which may simply activate a motor or perhaps begin a long, complicated subroutine.

2. Proportional signals, which might control the speed and direction of a motor.

3. Configuration signals, where an input control device is placed in a specific configuration by the human operator. The device then generates signals representative of this configuration and the teleoperator actuators try to attain the stipulated configuration. Often this kind of control is termed *position control*. It is employed in many master-slaves, exoskeletons, and walking machines.

4. Symbolic signals, with intrinsic, coded meaning.

Using the above classifications and the various types of physical controls associated with teleoperators, we can construct the matrix shown in

Table 5.2 Overview of Control Hardware

| Basic type of control | Hardware manifestations | Types of teleoperators using control type |
|---|---|---|
| On-off | Various switches (hand, eye, voice, muscle bulge, EMG) | Unilateral manipulators, artificial limbs, lunar surface samplers |
| Proportional | Potentiometers, joysticks, voice, muscle-bulge devices, EMG controls | Unilateral manipulators, artificial limbs |
| Configuration | Cables, potentiometers, servos (all located on the analog) | Bilateral manipulators, walking machines, exoskeletons |
| Symbolic | Typewriters, voice, punched cards, other software, switches | All teleoperators that employ supervisory control subroutines |

Table 5.2. Most of the more sophisticated teleoperators, it will be noticed, utilize configuration (position) control. In most cases, the control configurations are those taken by the human body; but this is not always true—analog controls may assume decidedly nonanthropomorphic shapes.

## Switches and Switchboxes

The simplest teleoperator control is the on-off switch. Many hundreds of unilateral manipulators now working in hot cells and on submersibles are controlled from switchboxes and switch consoles. The advantages of the switch are many: simplicity, low cost, reliability, no load reflected to tire the operator and switchboxes can be made small and portable, just the thing to carry from porthole to porthole within a cramped submersible. There are accompanying disadvantages, too. Switches are open-loop controls; there is no force feedback. Unless potentiometers or multiple-pole switches are used, there is no control over the rate of teleoperator joint movement. Only one joint can be activated at any instant during precision manipulation. (Some fast slewing motions can be carried out using more than one degree of freedom.) Lastly, switch arrays bear little resemblance to the manipulator configuration; they are decidedly nonanthropomorphic; and operator identification with the task is small. Nevertheless, in many applications, simplicity, reliability, compactness and low cost win out.

On-off switch control may be attained with toggle switches, push-button switches, or slide switches. If rate control is also required, a separate potentiometer can be connected in series with the switches. Often, however, proportional control of manipulator joint movement is achieved by installing rotary or linear potentiometers as the primary control elements. Pressure-sensitive resistance elements, strain guages, and piezoelectric elements can also provide an output proportional to the force applied by the operator. Proportional controls are usually spring-loaded so that they return to a null position when the operator removes his hand. Most on-off push-button and level-type switches also return to zero when released. Actually, manipulator joint motion is "three-valued," that is, *left/stop/right* or *counterclockwise/stop/clockwise*. The corresponding control switches are also three-valued. Three-way toggle switches (Fig. 5.10) are common and so are pairs of spring-loaded push buttons.

Switchboxes or control arrays are arrays of on-off switches, potentiometers, and feedback signals arranged in a convenient, logical fashion (Fig. 5.11) There will be one switch or pair of switches for each teleoperator degree of freedom. Switches may be color-coded; coding by shape is also common because the operator should keep his eyes on his work rather than the switchbox. To build in a little anthropomorphism, three-valued switches are connected to move a joint to the right when the switch is moved to the right and vice versa: ditto with up-and-down motion and rotary motion.

Switches and potentiometers usually connect directly with the electric motors that drive the joints in all-electric teleoperators. In electrohydraulic and electropneumatic teleoperators, switch-controlled electrical signals open and close valves (Fig. 5.12). When the teleoperator is far away from the control station, the control signals may be time-multiplexed, as is common in conventional remote control. In space work, the control signals may be digitized before transmission, as described below for the Surveyor surface sampler.

The NASA-JPL Surveyor surface samplers, while not particularly

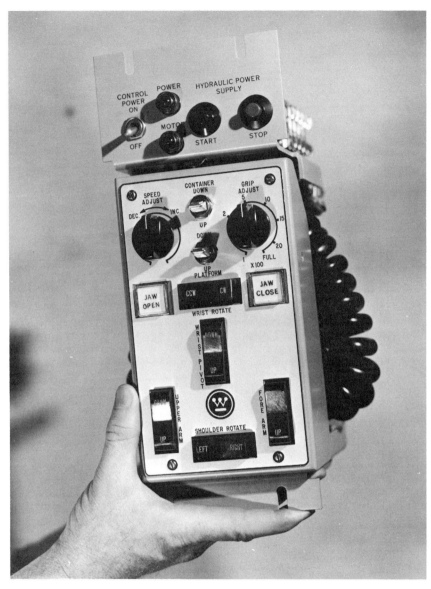

Figure 5.10 A control box for an underseas electrohydraulic unilateral manipulator. Most switches are three-way. (Courtesy of Westinghouse Electric Corp.)

114    The Control Subsystem

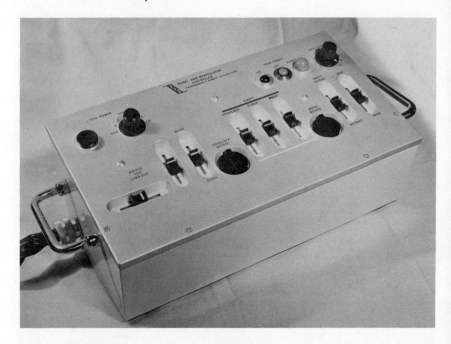

**Figure 5.11** Control box for the PaR Model–3000 unilateral manipulator. Contrast this switch-type control with the exoskeletal Handyman controls in fig. 32. (Courtesy of R. Karinen, Programmed and Remote Systems Corp.)

dexterous teleoperators, did carry out many lunar experiments during which they manipulated lunar soil and rocks. In one instance, a surface sampler was used to dislodge a Surveyor alpha-scattering experiment which was hung up on the spacecraft—a classic example of the use of a teleoperator for repair in a distant, hostile environment.

The sampler's four degrees of freedom were driven by reversible motors (Fig. 5.13) activated by digital commands dispatched from NASA's Goldstone Deep Space Network station in California. The only feedback to the operator consisted of television pictures (delayed by the signal transit time of about 1.3 sec) and telemetry signals indicating the current delivered to the motor being operated. Only one motor could be activated at a time—and then only in 2- or 0.1-second increments. Except for this quantization of motion, the surface sampler operated much like a unilateral manipulator in a terrestrial hot cell.

The sampler controls, however, were not the simple switches we associate with unilateral manipulators. The operator had to send digital commands to activate the proper motors. The selection of the digital

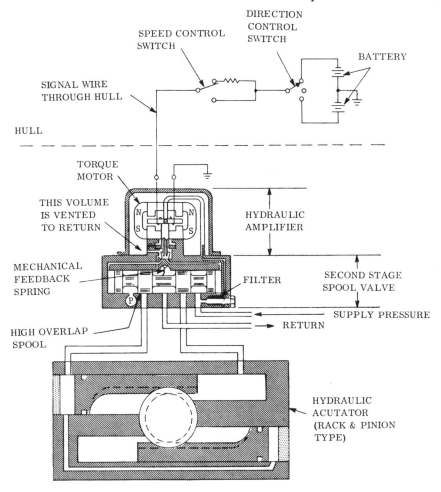

**Figure 5.12** A representative hydraulic pivot actuator employed in underseas manipulators. (Courtesy of Westinghouse Electric Corp.)

command word is analogous to selecting a switch on a switchbox and pressing it. The operator could also select, using additional commands, the length of the time the motor would run (2 or 0.1 sec) and the number of motion increments allowed. Thus, the operator could watch his television monitor and proceed stepwise through his experiment using the move-and-wait strategy recommended for time-delay situations.

Preprogrammed tapes were also employed for some sampler motions,

## 116 The Control Subsystem

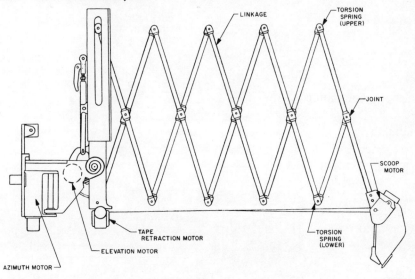

**Figure 5.13** The Surveyor surface sampler in extended position. There are only four degrees of freedom.

providing a form of supervisory control. This subject will be covered in more detail toward the end of the chapter.

### Joysticks

A joystick is a stick-like control that may be tilted forward and backward, sideways; it may also be twisted or pushed in and out along its axis. Buttons and switches are frequently mounted within reach of the operator's fingers while he is manipulating the stick. A joystick consolidates controls for several degrees of freedom into a single piece of hardware. Two important features of joystick control are proportionality (joystick displacement or pressure can be employed for rate control of a joint) and directionality (joint motion can be reversed simply by reversing the joystick polarity). The joystick illustrated in Fig. 5.14 shows how seven degrees of freedom can be controlled with a single joystick, although it is unusual to make joysticks so complex.

Joysticks may be either force-operated (*isometric* or "*stiff-stick*") or position-operated (*isotonic*). Manipulators have been constructed using both types. No clearcut advantages have been demonstrated for one over the other. Kelley has tabulated the relative advantages and disadvantages of the two types (Table 5.3).

A joystick is often a better control device than an array of switches because the operator identifies better with the task, particularly if the

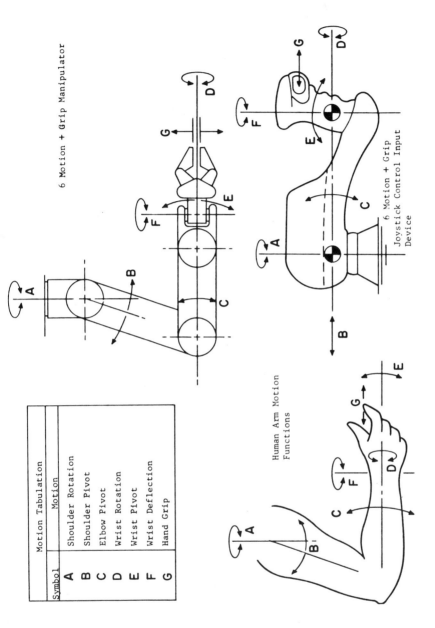

Figure 5.14 Three drawings showing the parallelisms between the operator arm, the joystick, and the remote manipulator. Despite the anthropomorphic character of the joystick, the manipulator is classified as "unilateral."

Table 5.3 Comparison of Force-Operated and Position-Operated Teleoperator Controls[a]

| Force-operated (isometric) | Position-operated (isotonic) |
|---|---|
| Controller output corresponds to forces applied by operator; "natural" control. | Controller output does not correspond to forces applied by operator; an interpretive step is required for control. |
| Controller output drops to zero unless manual force is maintained on the controller; i.e., it is self-centering. | Control lever remains at position last set; output remains applied without maintaining manual force. (Controller usually maintains a set position by virtue of sliding friction.) |
| A large output range may be accurately controlled by a small range of control lever displacement. | To control a large output range accurately, a large range of control lever movement is needed. |
| Large manual forces are required to control a large output range accurately. | A large output range can be controlled accurately with very small manual forces. |
| Because large manual forces are required to control a large output range accurately, a controller must be built and located so the operator may exert large manual forces on it. | Because a large output range can be controlled accurately with very small manual forces, many types of controls, in a large range of locations, may be employed. |

[a] Adapted from Kelley, 1963.

joystick is built along anthropomorphic lines like that pictured in Fig. 5.14. Crawford and Kama have compared operator performance on a unilateral rate-controlled manipulator using both a joystick and an array of levers (Crawford, 1961). They found the joystick to be superior. Pesch has compared the joystick against a pushbutton array and also found it superior (Pesch, 1967). The fact that several joints can be controlled by a single, rather anthropomorphic joystick also helps make this controller superior to switches and switchboxes.

The motions or pressures on a joystick activate the switches and (more commonly) potentiometers which control the joint motors. There is no force feedback in the usual joystick, although one can see how the addition of servos to the joystick might be accomplished.

Almost all unilateral manipulators now in operation employ switchbox controllers. But a few exceptions exist: one is the venerable General Mills 300 manipulator still in hot-cell use. Control of this unilateral manipulator is by two joysticks. The General Mills control console is a rather formidable piece of equipment, but most joysticks can be made smaller and more portable. Westinghouse, for example, has constructed a small joystick for controlling the electrohydraulic manipulator arm it built for the Deep Sea Rescue Vehicle (DSRV-1). The DSRV joystick has the feature of shifting from one degree of freedom to another in much the same way one shifts gears in an automobile. In the left-hand position, the operator twists the control one way or the other to activate the manipulator shoulder joint; in the right-hand position control shifts to the elbow pivot; the forward position takes care of the wrist pivot. The back position, however, activates a supervisory control subroutine called True Arm Extension; a straight in-and-out motion for scrubbing, sawing, etc.

## Analog Controls

Rather than push buttons or tilt joysticks to maneuver a manipulator arm into the desired position, why not make the controller an analog or replica of the working arm and design controls that force the working arm to duplicate the configuration of the control?* This, of course, is approximately what a master-slave manipulator does. Master-slaves are a step more complicated, though, because they also provide force feedback. The usual analog control displays no force feedback, although it could be built in as in the case of the joystick. In fact, an analog control can be thought of as a many-jointed joystick, although it is neither isotonic nor isometric.

Each joint in the analog control master arm has a potentiometer pickoff which supplies a signal to the real arm. If the slave arm is in a configuration different from that of the control arm, the corresponding signals from pickoffs on the slave arm will not correspond to those from the master control arm. Joint motors will be driven until all differences are nulled and master and slave arm configurations are identical.† The word "configuration" is used here intentionally because configurations but not linear motions are identical—even if the control arm is considerably smaller than the actual arm (a useful feature aboard a crowded submersible). During manipulator operation, an arm activated by an analog control may lag significantly behind the motion of the master

---

*This is called *analog control, replica control, model* (or *model-arm*) *control, position-servo control*, and, if the control arm is smaller than the real arm, *miniature-arm control*.
†The use of the words "master" and "slave" should not make the reader confuse analog-controlled manipulators with master-slaves, master-slaves can be actuated from the slave side (they are truly bilateral) but an analog-controlled unilateral manipulator cannot.

control because joint motors have limited speeds. Thus, there is not necessarily good spatial correspondence.

The advantages of analog control are several:

1. The controls can be made much smaller (or larger) than the slave arm/hand combination.

2. The controls can be activated from the master-hand area alone, with the rest of the joints following the hand motion naturally—like railroad cars. Some arms, such as those of Project MAC, are made with many more joints than the human arm to facilitate this sort of *terminus control*.

3. Force feedback can be accommodated easily.

4. Preprogramming for supervisory control consists simply of providing simulated pickoff voltages. As an alternate, routine motions can be accurately carried out using a grooved template; with terminus control the grooves are in essence preprogrammed instructions. (Picture the operator holding the master terminus like a pencil and following the template grooves.)

There are some very real difficulties, too:

1. Cost and complexity.

2. It is difficult to incorporate physically all the necessary circuits and electrical components, particularly in miniature master control arms.

3. With a miniature analog control, small motions are greatly amplified in the slave arm.

4. Considerable friction or an automatic braking system has to be built into the analog control arm so that it (and the slave arm) will not collapse when the operator releases it. This friction may fatigue the operator.

Almost all of the organizations engaged in undersea manipulator development have experimented with miniature analog controls. Generally speaking, the experiments have shown that analog control is feasible but that the disadvantages just listed outweigh the positive features. Most undersea manipulators are still controlled by switchboxes and joysticks.

## Master-Slave and Similar Bilateral Controls

The master-slave manipulator concept was pioneered by Argonne National Laboratory (ANL) in the late 1940s, when R. C. Goertz's group developed the first mechanical master-slaves. Later, ANL developed a series of electrical master-slaves. Master-slaves are bilateral teleoperators in which forces and torques at the master controls are proportionally reproduced at the slave actuators and vice versa. Normally, there are seven degrees of freedom, all of which can be controlled simultaneously. Master-slave controls have fingers, shoulder, and wrist joints, which make them considerably more anthropomorphic than the switchboxes and joysticks just discussed. Operation of master-slaves is natural and the operator easily projects himself into the work area. Spatial correspondence also exists. The mechanical master-slaves, particularly the famous ANL-developed Mod-8, are relatively inexpensive, reliable, versatile, and easy-to-operate. They are among the most common teleoperators in operational use.

The seven degrees of freedom in the mechanical master-slaves are activated by mechanical linkages that physically tie the controls to the actuators. Many of the operator's motions, say, wrist action, are communicated via metal tapes or cables. In a sense, the mechanical master-slave is a rather complicated pantograph, with one-to-one spatial correspondence. It differs from the analog controls in the sense that it is completely mechanical and possesses force feedback.

The master hands of the mechanical master-slaves are the focus of the operator's attention. When he wishes to pick up an object he moves the master hand to the object; both master and slave arms accommodate; this is much like the terminus control employed with analog controls. It is position control as opposed to the rate control employed in most switch and joystick-controlled unilateral manipulators. Once in the vicinity of the object, the operator makes fine position adjustments and orients the wrist. He then grasps the object with the fingers or tongs. Thus, master-slave manipulation actually consists of coarse terminus control followed by fine hand adjustments.

Master-slave hands are anthropomorphic in that the slave fingers are controlled by the human thumb and forefinger, the same digits we use to pick up objects in everyday life (Fig. 5.15). The wrist joint, too, is "natural." The typical master hand also possesses some joystick characteristics. The pistol grip is surrounded by switches, buttons, status lights, and levers that add versatility to the teleoperator. To illustrate, the manipulator may be locked in position so that the operator may leave his station without having master and slave collapse under the influence of gravity. Force amplification may be introduced between master and slave fingers. Even with these "unnatural" side benefits, master-slave controls represent a large step toward anthropomorphic controls.

The physical configurations of the ANL electrical master-slaves are patterned after their mechanical predecessors. The master hands, for example, are similar in both the mechanical and electrical species. The major difference between the two is that the metal tapes and cables connecting master to slave are replaced by servos and electrical signals. The

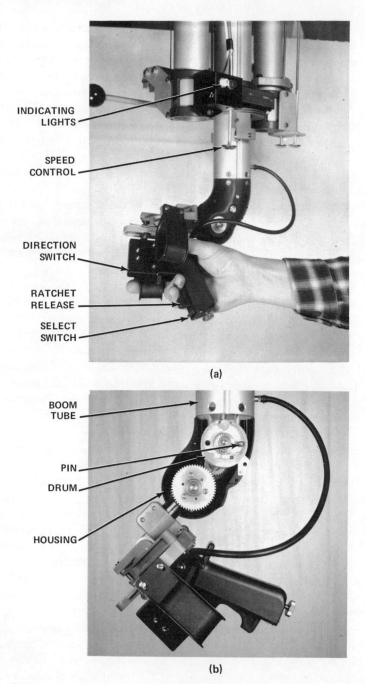

**Figure 5.15** Master hand of a Central Research Laboratories Mod-8 mechanical master-slave. (a) fully assembled hand, (b) hand disassembled, showing wrist gearing and tape drum. (Courtesy of Central Research Laboratories.)

electrical signals may move via hardwire or radio. It is this last fact that greatly increases the versatility of the electrical master-slave over its mechanical cousin. Master and slave stations can be hundreds of feet or millions of miles apart when one dispenses with mechanical linkages. Electrical master-slaves therefore can be considered for use in outer space or wherever great distance separates master and slave stations. Moreover, penetrations in hard-to-seal barriers, such as spaceship or submersible hulls, are easier to design for electrical wires than moving tapes and cables. With the added versatility of the electric master-slaves came increased cost and complexity. Because of these factors, electrical master-slaves have not yet been widely used.

In the ANL electrical master-slaves, the operator's input motions are first communicated to rotary drums with position sensors by means of metal-tape linkages. Thus, the master controls are similar to the mechanical master slaves up to the drums (Fig. 5.16). On the slave side, the

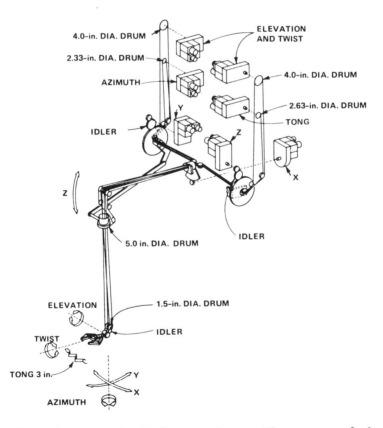

**Figure 5.16** The ANL Mark E4A slave-arm schematic. The master-arm displays a similar configuration.

situation is reversed; servo motors drive drums and metal tapes that actuate the corresponding degrees of freedom. Each of the seven degrees of freedom requires a master servo drive unit with two, 60-cycle, low inertia servo motors. As in all true master-slaves, the slave hand and arm can control the master—the real meaning of bilateralness. On the slave side, four servos are used. Geared synchromotors on each side provide

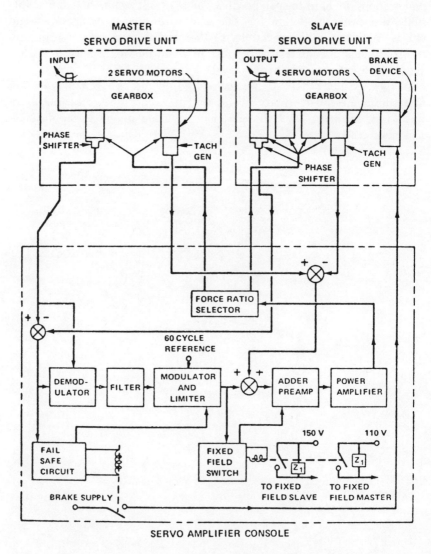

**Figure 5.17** Servo block diagram for the ANL E4A electrical master-slave (one degree of freedom only).

positional information. The servo system block diagram for the ANL Mark E4A is presented in Fig. 5.17.

During 1966 and 1967, ANL participated with NASA's Marshall Space Flight Center and their contractor, Ling-Temco-Vought, in a study of manipulators for use in orbital spacecraft (Argonne National Laboratory, 1967). One of the spacecraft studied was the Space Taxi, illustrated in Fig. 5.18 with two electrical master-slave working arms and three less

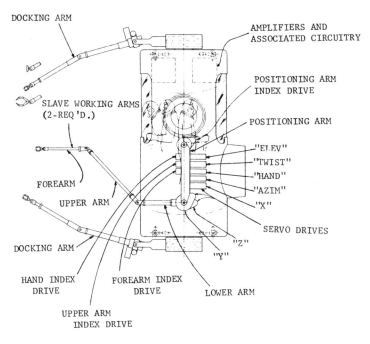

Figure 5.18 The ANL-LTV-MSFC Space Taxi master-slave electric manipulator arrangement. There are two master-slave working arms and three docking arms.

sophisticated docking arms. In this study, ANL proposed mechanical master-slaves for early availability and electrical master-slaves as the best solution, given adequate development time.

One of the basic control problems encountered in this study was the operator's limited working volume, a situation reminiscent of that on small submersibles, where switchboxes and joysticks are the common solutions. The ANL-recommended teleoperator configuration inverted the usual master-slave arrangement. The working arms are mounted at the spacecraft bottom, below the operator's feet, giving him an unobstructed view and freeing cabin volume for torso and arm movements. Instead of the usual master hand control (Fig. 5.15), a master handle or joystick with a trigger and switches was proposed. The master handle (the analog of the slave hand) would be position-controlled, just as in the normal terrestrial master-slave.

The problem of restricted operator volume was solved in the ANL study by the use of indexing. Indexing involves driving the slave arm independently of the master. If the operator cannot reach something because he has reached the limit of movement of the master control, he can gain additional slave arm motion through the use of indexing motors. As the slave moves, the master can be repositioned. The effective working volume of a bilateral master-slave can thus be increased by unilateral, switch-controlled motors. The indexing control switches are often located on the master hands in master-slaves. Obviously, there is some loss of spatial correspondence when indexing is employed—the price of expanding operating volume.

Automatic indexing appeared promising in the ANL space study. Whenever the master hand would reach the bounds of the operating envelope, index motors would automatically switch on until the master hand was operating again in the prescribed volume.

Several other electrical bilateral manipulators have been built. One of the most sophisticated and most interesting was the Handyman electrohydraulic bilateral manipulator built by General Electric for the Aircraft Nuclear Propulsion program in the 1950s. Handyman, with ten servoed joints in each arm-hand combination, was better articulated than the ANL seven-degree-of-freedom master-slave. Another departure from the basic ANL master-slave configuration was the more faithful paralleling of human joints with control joints. As shown in Fig. 5.19 the Handyman master controls are almost exoskeletal, particularly the forearms and hands. Even the hand is articulated. In other words, Handyman takes a further step toward anthropomorphism. One of the rewards is greater dexterity and more compliance with human manipulatory tasks.

At the AEC's Brookhaven National Laboratory (BNL) another electrical bilateral manipulator has been developed for use with high energy accelerators (Flatau, 1965). The BNL approach differs from that used by ANL in the application of D.C. servos at the joints themselves instead of at the top of the manipulator connected by tapes and cables to the joints. Flatau has claimed the following direct advantages:

1. Better frequency response due to direct coupling of motions.
2. Complete articulation of all motions, such as continuous rotation of the slave joints due to the absence of interconnecting cables.

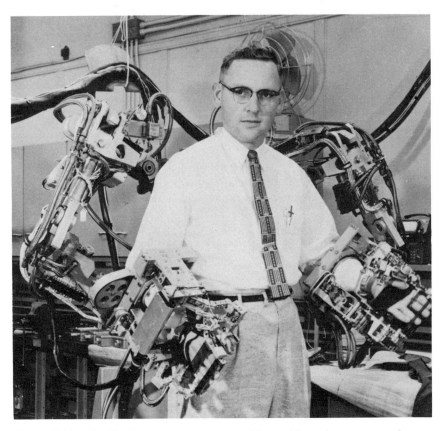

**Figure 5.19** The Handyman master station, with ten bilateral servos in each arm-hand combination. (Courtesy of R. S. Mosher, General Electric Co.)

3. Simplification due to the absence of metal tapes and cables.
4. Higher reliability (no cables, tapes).
5. Lighter and more compact.

Naturally, there are disadvantages, too:

1. Many different servo packages.
2. More complex servo schemes to reduce reflected friction and inertia.

Summarizing, the bilateral master-slaves and other associated bilateral manipulators add the dimension of force feedback to manipulation. Except in the case of the simple and ubiquitous mechanical master-slaves, the cost of mechanizing force feedback in terms of money, complexity, and reliability has militated against widespread application. The ANL electrical master-slaves have proven highly versatile and effective in hot-cell work, but still they have not been adopted for underseas work—a most logical application from many standpoints. When manipulation at great distance is contemplated, as in the space program, there is no alternative to electrical master-slaves if force feedback is considered essential. However, even here there is a catch: the greater the distance separating the master and slave stations, the greater the time delay in force feedback. While electrical master-slaves may be useful in orbital work, they may be far less attractive on the Moon because an Earth-based operator will not feel the reaction forces for over one second. However, teleoperators controlled from a manned lunar lander would be very useful in reconnoitering the Moon; time delay would be negligible here.

## Walking Machine Control

If good hand-arm teleoperators can be manufactured, why not leg-foot teleoperators; that is, a walking machine, a pedipulator rather than a manipulator? Walking machines can be made with ease; a great many of them have been constructed over the ages, from small walking toys to huge drag-line machines used in mining work. All of these machines have one thing in common; they are preprogrammed. Being preprogrammed they are completely deterministic, treading ahead blindly regardless of the terrain or obstacles. Even the more modern and sophisticated experimental walking machines built by Shigley and Space/General permit the operator to do no more than start, stop, and steer. For good off-road mobility, however, we need either a highly adaptive, operator-less control system or a human operator to provide the adaptation to terrain and obstacles in person.

Earlier we mentioned the digital, adaptive control scheme suggested by Hoch, at Battelle–Northwest Laboratories. In this scheme, adaptation to terrain is accomplished by the analysis of feedback signals from the joints of the walking machine. Because the operator would merely drive the propelled vehicle, the machine would not be a teleoperator as defined in this book. To qualify as a teleoperator, a walking machine must have man in the control loop, although he might initiate certain walking subroutines, particularly on easy, relatively flat terrain and when operating on prepared surfaces.

If man is to be an intimate part of the control loop, the first impulse is to build the walking machine like man; that is, a biped. Instead of controlling teleoperator arms, man would control legs. R. S. Mosher, at General Electric, has suggested such two-legged pedipulators. General Electric is now developing a man amplifier under DOD sponsorship which is essentially an exoskeletal biped walking machine with servoed arms. However, the purpose of this machine, which is called Hardiman, is man amplification, not off-road mobility, and we reserve discussion of this machine until the next section.

There seems little argument that a biped pedipulator would work if carefully controlled by man. General Electric has actually built a Pedipulator Balance Demonstrator that has proven that man can balance himself easily atop a two-legged servoed machine (Mosher, 1967). In operating this balancing machine, the operator's head is some fifteen feet from the floor and there is a natural fear of falling. Nevertheless, operators quickly learn to rely on their senses of balance and control the machine successfully. From the neuromuscular standpoint, a neophyte operator "knows" how to operate the machine immediately—the GE balance machine is that anthropomorphic.

Despite the success of the Pedipulator Balance Demonstrator, Bradley and others have pointed out that a biped walking machine can still fall, just as a man does on occasion, then the machine would be out of commission until a crane came along to right it (Bradley, 1967). For this and several other reasons not associated with control, development interest has now focused on quadruped walking machines. In this kind of teleoperator, the human operator controls one pair of legs with his legs and the other set with his arms—more or less as if he were crawling. Objections to the four-footed walking machine concept have been raised by engineers at the Army's Rock Island Arsenal (Rock Island Arsenal, 1968). Briefly, this critique asserts that walking stability in a quadruped is a strong function of its active torso. For example, no gait can be found that does not call for lifting a leg on a heavy corner; unless the animal's torso helps shift the center of gravity there will be a fall. Since the quadruped walking machine does not possess a controlled, articulated torso or other means of shifting balance, this critique claims that instability is likely. The Rock Island Arsenal report concludes that hexapeds or octapeds—controlled automatically—would be more reasonable engineering solutions to off-road mobility.

The quadruped concept is being tested in a General Electric development program sponsored jointly by the Army Tank-Automotive Command and the Advanced Research Projects Agency. The objectives of the Walking Truck or Quadruped program are to design, construct, and test a full-scale, four-legged walking machine capable of carrying an operator and 500 pounds of cargo. Each leg of the Walking Truck has three joints powered by force-reflecting hydraulic servos.

A major requirement in the Walking Truck program is the development of effective operator controls. A full-scale simulator was built to test out ideas. The simulator was unpowered but the controls were mechanically connected to the truck legs to provide force feedback and position spatial correspondence. During operation, the simulator was suspended by a crane, and the operator executed walking and turning maneuvers. Human factors analysis of simulator tests indicated that satisfactory control of all leg motions could be accomplished by a single operator. The simulator, of course, could not check out the assertion of the Rock Island Arsenal engineers that the machine would fall over in practice.

**Man-Amplifier Control**

A man amplifier is an exoskeletal teleoperator that greatly increases the physical strength of the operator wearing the structure. The artist's concept of the General Electric Hardiman readily shows the marriage of the Walking Truck legs (Fig. 5.20) with the Handyman arms (Fig. 5.9). The result is a machine envelope for man, with many but not nearly all of man's articulations copied with bilateral servos. The design and applications of man amplifiers are discussed in more detail in Chapter 2.

Although Cornell Aeronautical Laboratory did considerable exploratory work on man amplifiers in the early 1960s (Clark, 1962), the more recent General Electric Hardiman project is the only effort that has attacked the hardware problems in depth (General Electric, 1966, 1968). Hardiman, with fifteen degrees of freedom (some in series), is considerably more complicated than even the Walking Truck. The project should be classified as "exploratory hardware development."

In the Hardiman concept, the operator stands inside an anthropomorphic structure built in two halves that are joined together only at the hips by a transverse member called the "girdle." The exoskeleton parallels the operator everywhere save at the forearms, where the exoskeleton completely surrounds the operator, and his arms are colinear rather than parallel with the exoskeleton forearms. This forearm arrangement simplifies controls and makes it easier for the operator to identify his arm with the slave arm. The slave hand consists of one servoed degree of freedom that forces an opposed "thumb" toward a V-shaped palm-finger structure. An additional thumb-tip joint is not servoed but responds to an operator on-off switch control.

The force ratio contemplated between master and slave structures is about 25. This immediately raises a question of operator safety should the slave exoskeleton somehow run amok. In the GE design, limbs are physically linked in such a way that small master-slave errors cannot build up to do damage. Another safety feature locks all actuators should hydraulic pressures or control signals fail. Collapse of a heavy exoskeleton —carrying perhaps a 2,000-pound load—would be very hazardous without such a provision.

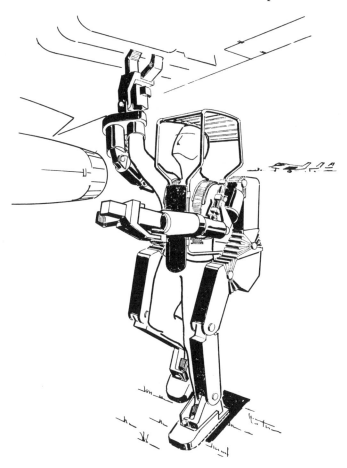

**Figure 5.20** Sketch of the General Electric powered exoskeleton concept (Hardiman I)

The articulation and dimensions of the GE man-amplifier were determined by a study of the motions that it could perform and the range of individual operators that it could accommodate without major adjustments. Operators were assumed to range from the 10th to the 90th percentile in physical size. Ultimately, the degrees of freedom and dimensions illustrated in Fig. 5.21 were selected for each side of the master-slave. With 15 joints on each side, a man-amplifier could carry out most of the important human motions, save for those requiring considerable dexterity of the hand.

In the original Hardiman concept, the operator exerted a force against the closely fitting control surface at any particular degree of freedom.

The surface then moved relative to the encasing slave member and, in doing so, actuated a valve in the master control circuit. Several schemes were proposed for translating the operator movements into signals that would actuate the hydraulically powered slave joints. One was a simple "tickler" or finger connected directly to the hydraulic valve. Tickler control was found to be unsatisfactory for the man-amplifier legs; and control of the joint angles was proposed for some leg degrees of freedom.

By early 1968 the mechanical design of Hardiman-I had progressed to the point where it was evident that a machine housing a human controller could be built that could lift and manipulate 1500 pounds. Mechanical-hydraulic bilateral servo development, however, had not progressed as rapidly as General Electric had expected. The key development problem concerned the stabilization of three or more servoed joints in series.

The Walking Truck and Handyman programs had proven that servo cascading was possible to a degree. But the Hardiman control requirements were so much more demanding that instability was likely using mechanical-hydraulic servos. General Electric, therefore, recommended replacing some mechanical-hydraulic servos with electrical-hydraulic servos because the latter can be stabilized rather easily using electrical circuits. Hardiman design is now proceeding on the basis of this change.

### Eye and Voice Controls

Although teleoperators are primarily manipulatory machines and normally should be controlled with the corresponding human extremities, there is no *a priori* barrier to the use of other parts of the operator's body for special control tasks. Man has no prehensile trunk or tail, but his eyes are remarkably well-controlled and, as we shall see shortly, his voice can be rich in symbolic commands.

It is difficult to mechanically harness the eye and derive control information from its motion. Optical pick-offs, however, have been developed for switching, gun aiming, and other purposes. Just how much of this technology is applicable to the teleoperator field?

NASA has developed an eye switch that depends upon the marked difference between the infrared reflection coefficient of the iris and the area surrounding it. The wearer-operator can voluntarily switch equipment on and off by directing his eye toward the infrared light source. As his eye moves, the infrared sensor mounted on the glasses frame detects the change in reflectivity and a switch is thereby closed or opened.

Honeywell and other organizations have designed and built eye-controls (called oculometers) that permit continuous control of machines (Merchant, 1967). In the oculometer the eye is illuminated with collimated light that is reflected by the cornea. The position of this reflection,

Teleoperator Controls 133

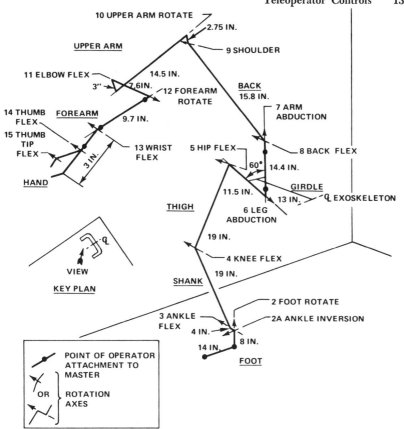

**Figure 5.21** Isometric stick figure showing the kinematic design of one half of the Hardiman exoskeleton. There are 15 degrees of freedom.

relative to the center of the pupil, is proportional to eye direction. To obtain a control signal, the pupil area of the eye is imaged into the photocathode of an image dissector tube. The pupil-iris boundary and the corneal reflection are acquired and tracked. The eye-direction control signal can be computed by comparing the relative positions of the pupil and the corneal reflection.

The primary applications of oculometers are in visual search, tracking, and instrument pointing. Conceivably, oculometer signals could steer walking machines and perhaps point sensors in the operating space, say, a television camera. In the next chapter, we shall see how head controls (not oculometers) have been employed to visually immobilize a TV scene in remote operations.

Another control signal, the human voice, can convey a great deal of abstract control information to the machine. For example, if a voice control were combined with the oculometer discussed above, the operator would merely look at an object and verbally command the machine to pick it up, turn it over, or move to the spot where his gaze is next fixed.

The most primitive kind of voice control depends only upon the presence of sound—any sound—to activate a switch. Such voice switches do not discriminate between natural noises and commands from someone other than the operator. Somewhat more selective are voice switches that depend upon a certain tone, perhaps a whistle. When the control system can discriminate between different tones (whistles, again) the operator could actuate several different switches or even continuously control the setting of a control. The grip force of a manipulator hand, for example, might be made proportional to the frequency of a whistle. Voice controls of this type have not been applied to teleoperators, although a few experimental devices have been constructed to help handicapped persons.

A single word in the human language can convey much more than an on-off switch command. Consider the fact that one human can verbally direct another to carry out the most complex task—in fact, any task that might be assigned to a teleoperator. Why, then, cannot a human operator verbally direct the machine portion of a teleoperator to carry out any manipulatory task he has in mind? In terms of tomorrow's technology, he probably can; but today's machines can comprehend only the simplest spoken words. Once they understand the words in a command, though, they can carry out the command to the letter. To illustrate, it is not too difficult to build a machine that can comprehend and act upon verbal "stop" and "go" commands.

Stanford Research Institute, M.I.T., and several other organizations are exploring the technology of machines that understand the spoken word. Sheridan's group at M.I.T. is the only one currently applying this approach to manipulator control (Sheridan, 1967). The basic manipulator commands are quite simple; at least at the lowest, most primitive level; viz., move wrist clockwise, close hand, etc. If this were not so, manipulators could not be controlled by simple switchboxes. The simple command "pick up," however, contains more information than that intrinsic in the flicking of a single switch. To pick an object up with a unilateral manipulator, several switches and their corresponding degrees of freedom may have to be activated. The human language moves easily from simple to complex commands, and today's machines are ready learners. The situation is analogous to the historical progression from the early machine-language programming of computers to the more and more abstract human-oriented instructions of Fortran and its descendants.

The first problem in voice control is, of course, speech recognition or the identification by the machine of the spoken word; that is, correlation of a train of sound patterns with known words in its memory. Once this association can be made, the computer can take over. The machine recognition of voice patterns is not within the scope of this survey. The reader should refer to the literature on the subject (Yilmaz, 1966; Peterson, 1966).

At M.I.T. an English-language-controlled manipulator is being built using a cascade of three processes:

1. A sentence parser, which recognizes typed (not spoken) words and casts them into categories, such as named objects, goals, specific actions, adverbs, etc.

2. A semantic interpreter which operates on the structured statement so that it can "understand," i.e., decide upon unique subgoals.

3. A manipulation interpreter, which, given the understood subgoal, decides upon a sequence of primitive manipulator actions to achieve that subgoal. This process may make use of state-space algorithms, or heuristic techniques, such as those mentioned earlier.

## Special Controls Used in Prosthetics and Orthotics

An artificial limb is much like a manipulator but the amputee who operates it is at a great disadvantage because he has either lost all or part of the analogous flesh-and-blood limb. The amputee often has recourse to his remaining hand for actuating controls or he may employ his shoulder, his feet, and other muscles. Even a person who is almost totally disabled can manage to move his head or tongue or some portion of his body to initiate externally powered aids.

If the artificial limb is externally powered, perhaps by batteries or compressed gas, switches located somewhere on the body are the most common sources of control signals. Switches are simple, cheap, and reliable —as mentioned at the beginning of this chapter. They are also very limited in flexibility and proportional control is impossible. However, an artificial limb can be controlled rather well with a few simple switches. A joint on a prosthesis is essentially a four-state device, just one step more complex than the ordinary fixed-rate unilateral manipulator joint. According to Tomovic, the four possible states are: (1) locked, (2) increasing, (3) decreasing, and (4) free (Tomovic, 1966). The final state is the one not found on ordinary unilateral manipulators. Proportional control is desirable in an artificial limb to improve manipulation and make its motion appear more natural. However, the simplest artificial limbs are generally the most successful because the ordinary wearer does not wish to be burdened with additional complexities.

The classical way to control artificial limbs is through shoulder shrugs and other bodily motions that pull cables attached directly to the artificial limb. The following general discussion applies to both self- and externally powered devices.

McLaurin has reviewed the different approaches to musculoskeletal control in prosthetics and orthotics (McLaurin, 1966). He classifies the control motions into three groups:

1. The motion of one part of the body relative to another; for example, shoulder elevation, chest expansion, chin movement, humeral flexion, elbow flexion, finger motion, and many, many more.

2. The motion of one part of the body with respect to a fixed object; for example, head motion relative to a wheelchair, torso motion (joystick-fashion) relative to a chair, and, of course, the eye controls introduced earlier.

3. The motion of parts of the body relative to space; for example, head motion relative to local gravity and head motion that causes gyros to generate a control signal.

Many harnesses and special cables have been devised to help an amputee control and actuate an artificial limb without external power (Anderson, 1958).

The cable that runs from the harness down to the arm is called a Bowden cable. On occasion, the control cables are surgically connected to the wearer's muscles in an operation called "cineplasty."

Wearer-actuated artificial limbs have been in use for centuries. But now that compact sources of power have been developed, interest has turned to the so-called externally powered artificial limbs and orthotic devices. The most common sources of power are electrical, pneumatic, and hydraulic. If the wearer of the prosthesis has one good arm, the requisite switches or valves may be located in his pocket or attached to his body where they are readily accessible. Switches have sometimes been located in the shoe, when the wearer did not want to be too obvious in controlling his prosthesis. Muscle-bulge switches are also employed, but like the shoe switches these control only one degree of freedom or one speed unless logic circuitry is added that translates coded switch signals into more sophisticated motion; i.e., two pulses, slow; three pulses, fast; etc. Each controllable degree of freedom might have a digital address; three- or four-level commands might be transmitted, too. But such codes are generally too much trouble. A few simple on-off switches are the rule.

When both arms are shrunken and deformed, as they are in many thalidomide cases, special switchboxes can be installed where they can

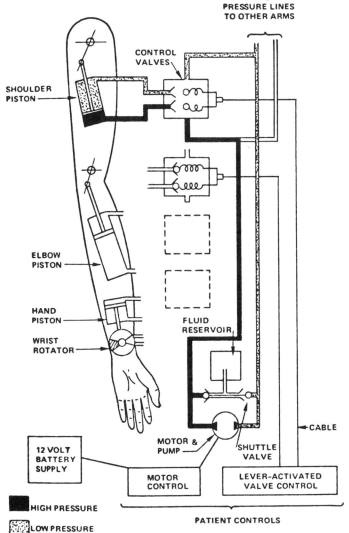

**Figure 5.22** Control system for the Northern Electric hydraulic arm. (Courtesy of Northern Electric Co., Ltd.)

be manipulated by the deformed limb and hand (the phocomelic digits) (Fig. 5.22) (Stevenson, 1967).

When the hands cannot be employed at all for switch control, the tongue turns out to be surprisingly responsive and effective. Many different types have been built. A seven-lever tongue control has been built by Rancho Los Amigos Hospital for high-level paralytics confined to wheelchairs (Karchak, 1968). Each switch has three positions. Much more elaborate *joystick-like* tongue-control devices have been designed wherein the tongue "manipulates" various levers and buttons. Proportional controls have been built into some of these devices.

The switches and proportional controls activated by muscle bulges are of three basic types: carbon, photoelectric, and strain-gauge (Lucaccini, 1967). The carbon transducers operate on the same principle as telephone transmitters; carbon granules are sandwiched between two electrodes. Muscle pressure on the electrodes will decrease the electrical resistance between the electrodes. Photoelectric transducers can be made in several configurations.

The fundamental idea here is to reduce (or increase) the quantity of light received by the photocell as the actuating muscle is flexed. Strain gauges can be attached to muscles to yield a signal roughly proportional to the muscle bulge.

Except for the artificial arms that are controlled directly by cables connected to body harnesses or actual muscles, there is little or no force feedback present in the schemes discussed above. Neither is there an analogous limb that would be able to interpret the feedback in most cases. This is a severe handicap because objects often fall from the grasp or are perhaps broken when the prosthesis wearer cannot feel the force he exerts. This deficiency can be compensated for to some extent by building a closed-loop control circuit that bypasses the operator. Salisbury *et al.*, at Walter Reed Army Medical Center, have installed piezoelectric sensors in the fingers of an artificial hand. These sensors detect the vibrations created when two surfaces slide over one another. Slippage noises are converted into commands that cause the hand grip to increase until slippage stops.

### Electromyographic (EMG) Control

Muscle activity is basically electrical in nature. When electrodes are attached near or in any of man's striated muscles, muscle flexure generates electrical signals we can pick off for control purposes. These signals are variously termed electromyographic (EMG), or myographic, or muscle-action potentials (MAPs).

Most of the work described below was carried out with the application to prosthetic and orthotic devices in mind. However, normal people generate EMG signals, and these may eventually be employed for controlling other kinds of teleoperators. Tiny electrodes, for example, may turn out to be much smaller and more comfortable to use than the controls described in the preceding sections. One can even visualize gloves or tightly fitting jackets, even space suits, with built-in electrodes that an operator

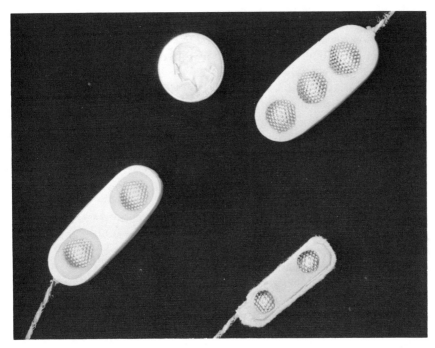

**Figure 5.23** Surface electrodes for detecting EMG signals. (Courtesy of W. Waring, Rancho Los Amigos Hospital.)

would don to control a teleoperator with many degrees of freedom. In this concept, the motions of the operator would be faithfully duplicated by the actuators, located perhaps in a hot cell or on the Moon. Such visions are far off, however, for the EMG state of the art is still rather primitive.

There are three classes of electrodes which may be used to pick up EMG signals: skin-surface types, types which pierce the skin, and types completely implanted in the body. Surface electrodes are obviously the easiest to install and remove (Fig. 5.23). Their disadvantages include weak signals—due to the high impedance of the skin, which can be reduced somewhat by electrode pastes—and the surface electrode's tendency to shift, producing "artifacts"; that is, unwanted electrical disturbances. There is also "crosstalk" from nearby muscles. Electrodes that pierce the skin may be placed just below the skin (subcutaneous) or they may penetrate the muscle itself. The skin impedance problem is bypassed by this kind of electrode. The intramuscular electrodes can pick off signals from different muscles or even from different parts of the same muscle. Electrodes that penetrate the skin still have a tendency to wander a bit; they may break off, too, leaving a piece of metal in the operator. Further, the electrode site is a source of irritation and potential infection. Surgically or hypodermically implanted electrodes that will survive the bath of corrosive body fluids and not irritate the operator are difficult to design. Surface and skin-piercing electrodes have been employed most frequently in EMG work.

All of the striated muscles are potential sites for control electrodes. In the case of the normal person, the muscle selected would ordinarily be analogous to the motor driving the same degree freedom in the teleoperator; the biceps, for example, might control a manipulator elbow joint. But almost any muscle can be trained for EMG-control purposes. Shoulder muscles are used for controlling artificial arms in cases where the natural muscle site no longer exists.* The teleoperator designer of the future may wish to seize upon this attribute for the control of nonanthropomorphic degrees of freedom; say, the control of wrist extension via a shoulder-muscle electrode. Even more exciting is the discovery that a human operator can voluntarily control single motor units in a muscle, a fact that potentially increases the number of available EMG control sites by a large factor. In other words, the number of output signals under conscious, voluntary control of the operator can be many times greater than the number of physical degrees of freedom. Despite these promises of future enhanced control through EMG, contemporary development programs are oriented toward making simpler systems work well, especially those destined for handicapped persons.

Let us look more closely at the EMG signals; when a muscle is flexed, electrodes nearby or embedded in the muscle itself record the summation of separate fiber action potentials (Wagman, 1966). In this sense, an EMG signal is an "interference pattern" resulting from the addition of numerous signals from separate fibers. The observed signal obviously depends upon the location of the electrodes. In spite of all the variables, normal muscles produce characteristic signal patterns. Three parameters describe these signals: amplitude, spike width, and spike frequency. Amplitudes are usually less than 50 millivolts peak-to-peak; while the spike width is measured in milliseconds. Spike frequency or repetition rate varies greatly with the muscle selected. A plot of power vs. frequency (the signal spectrum) is shown in Fig. 5.24 for two different muscle loads. It is the change in signal amplitude with muscle load that allows us to provide proportional EMG control to teleoperators. As muscle contraction increases, there is also an overall increase in repetition rate; another potential source of control data.

---

* In controlling orthotic devices, the natural muscle may still produce useful EMG signals even though the real arm cannot be moved voluntarily.

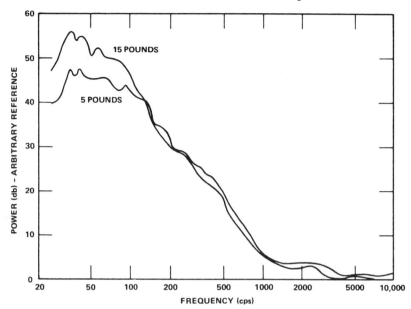

Figure 5.24 Typical power spectra for surface-electrode EMG signals.

Superficially, EMG signals would sound almost ideal for teleoperator control. EMG signals are certainly more convenient than, say, tongue switches for a handicapped person. They are also lighter and more comfortable than restrictive harnesses. Furthermore, reasonably accurate proportional control has been demonstrated. On the other hand, electric razors and other appliances may seriously interfere with EMG signals as does the crosstalk from other muscles. Some wearers of EMG-controlled devices feel that EMG offers less "positive" control than switches. Lyman, who has made a systematic study of the performance of EMG systems in skilled manual control tasks, found that the operators were easily fatigued and that considerable concentration was required, particularly when more than one degree of freedom was being controlled (Lyman, 1966). The poor reliability of bioamplifiers has also been a major problem area. Summarizing, practical EMG control is beset with development problems.

### Hardware and Software for Supervisory Control

Control equipment in supervisory control consists of both hardware (typewriters, computer consoles, and devices like light pens) and software (computer programs, tapes, analog records, etc.). Supervisory control hardware and software are in the experimental stage today.

Two groups have developed computer-controlled manipulators:

## 142  The Control Subsystem

1. Case Western Reserve, where NASA and the AEC have sponsored work potentially leading to the semiautomatic dissembly of radioactive nuclear rocket engines (Beckett, 1967; Taylor, 1966).

2. M.I.T. where general man-machine control problems are being attacked in both Sheridan's group (NASA-DOD support), (Barber, 1967) and Project MAC (DOD support) (MIT, 1963).

A typical system configuration of the Case computer controlled manipulator is illustrated in Fig. 5.25. The human operator can make inputs at two spots: the teletypewriter (TTY) and the conventional manual

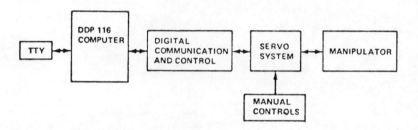

**Figure 5.25** Block diagram of the Case computer-controlled manipulator. The General Purpose Interface (GPI) described in the text bridges the interface between the computer and the manipulator.

controls. The TTY, of course, is the input point for supervisory control instructions, which may be extremely simple incremental motion instructions or a command to carry out a complex subroutine of instructions stored in the memory.

Following the path of supervisory control information—horizontally from TTY to Manipulator in Fig. 5.25—we see that the operator must first know what sort of instruction to type for the computer on the TTY; next, the computer output must be converted into electrical voltage signals that will drive the manipulator joints the proper distances in the proper directions. The interface between the operator and computer is bridged by the Teletype Executive Control Program (TTY Exec); while the computer directs the manipulator through the General Purpose Interface (GPI) Unit.

Supervisory control reaches its ultimate development in the fully developed robot: that is, a machine that carries out high-level, abstract commands without human assistance at any primitive levels of action. The robot developed by Stanford Research Institute (SRI) under DOD sponsorship (Rosen, 1968) is not a teleoperator per se because it manipulates nothing and rolls rather than walks (pedipulates) (Fig. 5.26).

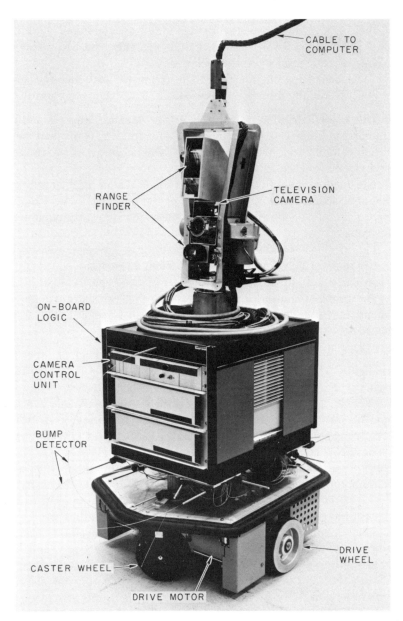

**Figure 5.26** The SRI robot. While this robot does not have the arms and legs needed to qualify as a teleoperator, its autonomous functions may eventually be incorporated into teleoperators. (Courtesy of C. Rosen, Stanford Research Institute.)

Nevertheless, arms and hands *could* be added and operated within the general hardware/software framework. In actuality, the SRI robot is a harbinger of future teleoperator technology.

The SRI scientists suggest that their robot (or any robot/teleoperator) will eventually be able to operate at four levels of control:

1. The immediate-action level, where the operator directly activates the motors and sensors. (This mode of operation is equivalent to the operation of unilateral manipulators by a switch box. In other words, no supervisory control exists.)
2. The tactical level, where the robot solves simple problems in navigation and locomotion without the help of the operator. (The VECTOR subroutine employed by the Case computer-controlled manipulator falls in this category.)
3. The strategic level, where the robot finds specified objects and relocates them. (The Case and M.I.T. computer-controlled manipulators can carry out supervisory instructions of this type.)
4. The problem-analysis level, where the robot translates a high-level command into a series of subtasks according to some criterion of performance.

The SRI robot is controlled through a teletypewriter, just as the Case manipulator. There is also an analogous computer plus its software (programs and subroutines). The computer-robot interface is intimate and specialized, not general-purpose like the Case GPI. While the Case system has status indicators, it does not have the full array of kinesthetic sensors possessed by the SRI robot. The SRI robot uses both the visual and kinesthetic feedback in local control loops.

Using visual feedback through its TV and kinesthetic data (obtained from bumping objects in its environment) the robot can construct a model of its environment. The model includes its own location as well as the positions, orientations, and in some cases identities of the objects. The robot can reconnoiter its surroundings itself—a valuable property for a teleoperator operating, say, beneath the sea, where feedback to the operator may be sparse. In most computer-controlled manipulators, obstacles to be avoided must have their coordinates placed in the computer's memory by the operator; not so with the SRI robot.

Clearly, by building upon a foundation of simple autonomous (primitive) functions and utilizing its feedback data, the SRI robot and its descendants will be able to carry out more and more generalized, high-level commands from the human operator. As stressed frequently in earlier chapters, teleoperators are also following this path toward greater autonomy.

# VI

# THE SENSOR SUBSYSTEM

Teleoperator sensors vary as much as the arms and hands we have described—perhaps more so, because man's sensors are more diverse and subtle than his extremities. At one sensory extreme, direct vision is augmented by a crude sense of touch conveyed through the handles of simple manipulator tongs; at the other, robots far from Earth may be controlled by an operator surrounded by banks of blinking electronic consoles and displays that convey to him the sight, sound, and feel of the alien environment. In teleoperator terminology, the operator wants to "project his presence" into a hostile environment or across distance. The function of the sensor subsystem is to reproduce faithfully those physical properties of the working space that the operator needs to do his job well. It does not mean duplicating *all* the color and thermal nuances of the environment—just enough sensations to carry out the required manipulations expeditiously. Even with this narrowing of the sensory spectrum, the engineering task is difficult.

Although this chapter deals with only three of the five categories of sensory feedback an operator may receive from the communication subsystem, all five categories of sensors are listed below:

1. *Vehicle navigation sensors.* This category includes gyros, loran, startrackers, direct vision, radars, and the myriad of navigation aids developed to pinpoint something (the teleoperator) in space, on land, or under the sea.

2. *Target tracking sensors.* Here, we include human eyes, TV, imaging sonars, radars, proximity devices, touch sensors, force feedback, and all sensors that tell the operator the position, velocity, and orientation of the target with respect to the arms and hands of the teleoperator.

3. *Target intrinsic-property sensors.* This rather unusual class of sensors conveys information about the weight, texture, hardness, temperature, radiation level, and other target properties that are independent of those sensed in Category 2.

4. *Teleoperator status sensors.* These sensors relay data about the health of the teleoperator, such as internal temperatures and summaries of switch positions. Also included are critical data telling the operator the positions, velocities, and accelerations of all degrees of freedom in the actuator subsystem. Commonly, the operator's eye receives such information directly or by TV. If the operator cannot see the scene, transducers on the manipulator joints may relay vital information.

5. *Environment sensors.* Teleoperators are more effective if the operators know something about the environment in which the target and manipulators are immersed. Microphones, thermometers, radiation detectors, ocean-current meters, and a wide spectrum of other "environment" instruments have been developed during nuclear, space, and underseas programs.

Most sensor possibilities in Categories 1 and 5 are already well treated in the aerospace, nuclear, and underseas literature. We shall concentrate upon those sensors that augment the operator's visual, auditory, and tactile senses, with only brief forays into more advanced concepts.

Table 6.1 Human Senses of Actual and Potential Interest

| Sensor category | Sense | Utility in teleoperators |
|---|---|---|
| Photoreceptors | Sight | Estimates distances, velocities, color, texture, orientation, etc. |
| Mechanoreceptors | Feel | Estimates weight, pressure, vibration, wind speed, impact, slippage, texture, size, etc. |
| | Hearing | Detects sliding, mechanical strains, motion, liquid flow, relay action, breakage, etc. |
| Chemoreceptors | Smell | Estimates composition, chemical reactions (fires), etc. |
| | Taste | Estimates chemical composition. |
| Thermoreceptors | Heat | Estimates temperature crudely, locates heat sources. |
| Electromagnetic receptors | | None known for man, though some fishes and (perhaps) birds use such senses. |
| Balance | Balance | Determines gravitational stability or the lack of it. |

Table 6.1 shows that men have more senses than the classical five, but manipulation is done mainly by sight and feel. The first task of the designer is partially to reproduce these two factors within the given cost and engineering constraints.

Auditory and visual alarm cues are useful, as subsequent sections will show, but without question major attention must be directed toward translating the physical situation in the working space into those sensory terms that the operator employs in eating, writing, and playing dominoes. Even nonanthropomorphic sensors, such as imaging sonars and precision radars, should render their information visually for human comprehension. In other words, the sensory picture should resemble scenes in ordinary life.

Each environment has its own sensory problems. Some are generic, such as adequate target illumination (light, sound, radar) and visual obstructions of the target by barriers and the manipulators themselves.

Table 6.2  Some Sensory Problems by Application Area

| Application area | Specific problems | Some solutions |
| --- | --- | --- |
| Aerospace | Limited communication bandwidth | Data compression, supervisory control |
| | Micrometeoroid damage to optics | Removable lens covers |
| | Ultraviolet browning of optics | Filters, special glasses |
| | Degradation of sensors during heat sterilization | Special heat-resistant electronics and sensors |
| | Vacuum welding | Special lubricants and materials |
| Undersea | Scattering of light by suspended particulate matter | Use long wavelengths and proper angle between light source and sensor. Imaging sonar |
| | Corrosion, fouling | Paints, coatings, remote cleaning devices, proper materials choice |
| Nuclear | Radiation damage to optics, electronics, and many sensors | Shields (perhaps movable), selection of radiation-resistant parts |
| Metal processing | Thermal degradation | Thermal shields, heat-resistant parts |
| Construction and mining | Explosion (shock degradation of sensors) | Shock absorbers, rugged construction |
| Public services | Thermal degradation (fires) | Thermal shields, heat-resistant parts |
| Entertainment | Concealment | Miniaturization of sensors, camouflage |

Others, such as light scattered by the sediments stirred up around a distressed submarine, are highly specific. Some other specific problems are listed in Table 6.2 along with some typical solutions.

## DIRECT-VISION SITUATIONS

Most of the teleoperators now in use let their human operators see the targets, the manipulators, and their spatial relationships directly. Every possible effort is made to improve the clarity, realism, and field of view available to the operator, because the eye represents the largest sensory input channel to the brain.

Direct viewing of the working space requires: (1) good lighting and (2) a good window or light path that lets the operator see without strain the light reflected from the targets and manipulator arms and hands. Lighting and light paths are not independent design problems. The intensity of light reaching the eyes of the operator depends upon both the intensity of the light source and the attenuation suffered in the window or other transmission media. Other design ingredients of somewhat lesser importance are distance, contrast, coloration, and arrangement of the targets with respect to the teleoperator hands and arms.

Lighting in outer space is notoriously variable, particularly on an Earth-orbiting, spinning spacecraft. In Earth orbit, any combination of the following situations may occur (Hyman, 1963):

| | |
|---|---|
| Solar illuminance | 13,500 ft-candles (lumens/ft$^2$) |
| Earthshine illuminance | 4500–9000 |
| Lunar illuminance | 0.03 |
| Starshine illuminance | 0.0001 |

Direct sunlight can render the target much too bright and require an astronaut to use his helmet's sunshade. If work is shadowed so that neither the Sun nor Earth illuminate it, the astronaut may not see it at all because his eyes will be adapted to the bright areas surrounding the work. Obviously, supplementary illumination that can be varied at will by the astronaut is needed (Ling-Temco-Vought, 1966).

Human-factors engineers generally recommend an illumination of about 3 ft-candles for fine work. To attain something close to this value, an astronaut may switch floodlights on when the work rotates into deep shadows and pull down his sunshade when the work is lit by direct sunlight. Intense contrasts can be softened by requiring spacecraft parts in peripheral areas to have light-absorbing, glossless surfaces. By deployment of nonspecular reflectors at strategic spots, some of the surplus sunlight and Earthshine can be reflected into dark regions.

Far under the sea, there may be no light save that provided by the vehicle itself, but the sea is almost as fickle as outer space in its perturbations of man's activities. For example, the attenuation length for 465-mµ light, five fathoms deep in the Atlantic off Gibraltar, is about 20 yards; off the Galapagos, the attenuation length is only about one-fifth this value. To add to the difficulties on the ocean floor, enough sediment may be stirred up by a submersible to make seeing considerably worse. Variable floodlights are needed.

Several concerns manufacture underwater lighting equipment (North American Aviation, 1966). Edgerton, Germeshausen, and Grier, for example, makes a series of quartz-iodine incandescent lamps, within quartz envelopes, that can be operated at depths down to 39,000 feet. Mercury-vapor lamps are useful, since their light is emitted in that portion of the spectrum where seawater attenuates light the least. Small manipulator-carrying submersibles mount several such lights at various points around the hull. A major problem with undersea (and space) lighting is the high power consumption of the lamps.

In hot cells and most other terrestrial applications there is no power-supply problem because power lines go nearly everywhere. Hot-cell interiors ar usually decorated with a glossless (flat) paint that reflects into all nooks and crannies, providing manipulator operators with almost ideal lighting conditions. The only major attentuator of light is the hot-cell window.

The viewing window of the hot cell, spacecraft, or submersible is an integral part of the communication subsystem. The overwhelming bulk of feedback information travels this route. Distortion, aberrations, and restrictions to the operator's vision must be eliminated as far as funding and technology permit. The main concern in space and undersea work is the integrity of the window in the vacuum of space on one hand and the crushing pressure of seawater on the other. Conditions are quite different in the nuclear field, where the hot-cell window must allow a man to see through a very thick biological radiation shield.

Periscopes and systems of mirrors gave early manipulator operators an over-the-wall peek at what was transpiring in hot cells. But these were tiring to use. In the late 1940's, Oak Ridge National Laboratory installed some small circular cylinders filled with transparent zinc bromide (specific gravity, 2.5) in the walls of hot cells. The zinc bromide attenuated the gamma radiation and still gave the operator a direct look at his work. Unhappily, these early windows were expensive and their transparencies deteriorated under large doses of radiation. The operator also had the feeling of "tunnel vision" with the small apertures.

Subsequent chemical research showed that the zinc bromide filling the windows could be stabilized in a radiation field by the addition of a reducing agent, hydroxylamine hydrochloride, in a concentration of about 0.01 percent. Windows were next widened to give the operator a better view of the hot-cell interior. Today, window apertures up to five feet square are feasible. Such dimensions give the viewer a field of view approaching 180°. Though liquid-filled windows may be as thick as five feet, the manipulator operator can project his "presence" readily into a well-lit hot-cell interior. The scene usually has a greenish cast (because of the window glass rather than the zinc bromide), but it is quite vivid. Sodium-vapor lamps, which emit nearly monochromatic light, prevent color fringes around objects in the cell. Distortion, which makes plane surfaces appear curved, becomes noticeable only when the viewing distance approximates the window thickness or when the viewing angle of incidence is greater than roughly 60° (Argonne National Laboratory, 1952).

Water has occasionally been used instead of zinc bromide to fill liquid windows, but its lower density makes it a much poorer radiation shield. Other dense liquids that have been tried include lead acetate, zinc chloride, and methylene bromide. With further development, some such fluids may approach the effectiveness of zinc bromide. Until this happens, zinc bromide dominates the field.

Some solid glasses are considerably denser than zinc bromide. Why not substitute solid glass plates for the fluid zinc bromide? The glasses available during the early atomic energy work were unstable in the presence of gamma radiation; they discolored or lost their transparencies quickly. The addition of cerium and other chemicals to the melt improved the situation markedly. As a result, one now finds some hot-cell windows constructed from several thick slabs of glass, as illustrated in Fig. 6.1. The spaces between the glass slabs are generally filled with mineral oil because its optical properties are similar to those of the glasses.

To give the operator a greater field of view, yet keep hot-cell wall penetrations down to reasonable sizes, the windows are frequently flared or stepped; that is, they open up toward the inside of the hot cell, just the reverse of a safe door.

Pressure takes the place of gamma radiation in fixing the sizes and compositions of submersible viewports. The deeper the submersible goes, the smaller and thicker its viewports. The North American *Beaver*, for instance, is designed for continental-shelf operation where pressures are not extreme. The *Beaver*, therefore, can afford a large panaromic, plastic window. In contrast, the deep submersible, *Alvin I*, designed for 6000 feet, can tolerate only small hull penetrations. Its plexiglass windows are

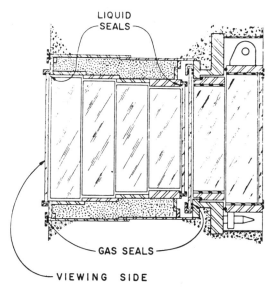

Figure 6.1 Sketch of the all-glass window installed in the Experimental Breeder Reactor II Fuel-Cycle Facility In Idaho. (Courtesy of Argonne National Laboratory.)

3.5 inches thick and 5 inches in diameter on the operator's side. The *Alvin I* windows open up conically at a 45° angle (Mavor, 1966). With this face close to an *Alvin* viewport, a manipulator operator would have a fairly large field of view, but there would be little room left for manipulator controls unless they were the small replica types (mentioned in Chapter 5) or switches.

In designing a viewing system, the engineer encounters several human-factors problems. The intensity of illumination and the spacing of flood-lights certainly falls into this category. Assuming good lighting, where should the work be placed relative to the operator (or vice versa, if one has no control over the work)? Should the work be color-coded? Will matching the work and the viewing system to the human operator improve overall teleoperator performance?

The human factors experiments of the Air Force's Aerospace Medical Research Laboratories provide some partial answers to these questions (Kama, 1964). For instance, the distance from the manipulator operator's unaided eyes to the work should not be greater than about 10 ft, less if possible. As distance increases, visual resolution and depth perception drop off and task performance time rises. The Air Force studies showed significant performance deterioration between 7 and 11 feet. Most manipulator installations, whether in the nuclear or underseas fields, provide the operator with mounted binoculars or telescopes to augment his vision should distances become too great. With visual augmentation, the distance to the work may be several times the recommended 10-foot limit, as it is in cavernous hot cells, such as E-MAD.

Wright Field studies also indicate that the work should be below the operator's horizon, i.e., he should look down on it. Task performance times were best when the angle below the horizontal was between 45° and 65°.

The loss of depth perception is a major factor causing deterioration of performance with distance. Air Force tests comparing binoculars and monoculars (telescopes) indicate that distant tasks are completed faster with the stereoscopic effects provided by binoculars. But contemporary, rather primitive, 3D TV apparently offers little if any advantage over 2D TV in manipulatory tasks. The two images needed in 3D TV are difficult to keep registered, especially if the cameras have to be redirected frequently to various parts of the work. Color TV, however, *may* improve task performance if the hands and work are specially colored to improve contrast. Further development may turn 3D TV into a superior viewing subsystem.

## VIEWING WITH MIRRORS AND FIBERSCOPES

Optical devices can help an operator see around a target or barrier:

1. When he cannot maneuver his vehicle or direct-viewing equipment into good positions.
2. When there are no viewports or windows in the barrier near the operator. Sometimes the radiation levels do not permit a man to work near barrier penetrations.

Mirrors are often placed strategically within cells so that hidden parts of the target may be seen from the window. Periscopes with their high quality optics are excellent for photography and magnified views of a target. Mirror and periscope viewing systems can be easily converted from fixed to scanning configurations by the incorporation of scanning mirrors.

How does one inspect remotely the inside of a tube or look inside a hatch? The borescope permits the inspection of pipes inside nuclear reactors and various other sites inaccessible to direct viewing. The borescope is essentially a periscope with its own light source. It comes with ready-made extensions that can effectively transport the operator's eye as far as 50 feet down a coolant pipe. An underwater oil well casing might be searched for a lost tool similarly; when the tool was located, a manipulator could recover it. This kind of application is similar to the Atomic Energy Commission's Down-Hole Project in which TV and a manipulator are combined to locate objects at the bottoms of deep holes drilled for underground nuclear tests.

The development of fiber optics has given the teleoperator another tool for examining inaccessible spots. A flexible bundle of hairlike glass fibers not only can carry light to the desired spot but also can bring out an image of the area. These so-called Fiberscopes have been made as long as 10 feet, and without question can be made longer. The remote manipulator would have to transport the viewing head of the Fiberscope into the target area. If the head can be secured, the manipulator operator can use its images to guide his manipulations.

## REMOTE TELEVISION

The practical utility of television in manipulator operations has been controversial for almost two decades. Some operators prefer TV to direct vision; others have no use for it. In some applications, nevertheless, TV is clearly superior, even mandatory:

1. In controlling a teleoperator at distances beyond the range of direct-vision optical equipment, viz., on Mars, etc.
2. In nuclear accidents when radiation precludes the close approach of men with direct-viewing devices.
3. In locations where there are obstructions to direct viewing, such as an undersea disaster area or a cluttered hot-cell floor.
4. In situations where simultaneous observations from widely separated various vantage points are required.
5. In situations when very little light is available.

Inherent in the above statements are the major TV advantages of portability, the ability to work under very poor lighting conditions, to use a wide selection of lenses (including zoom lenses), and the ability to focus remotely, change aperture, insert filters, and so on. Nevertheless, there are some drawbacks to the use of TV: cable-handling problems, very large bandwidth requirements, electronic instability, sensitivity to high radiation fields and intense light, limited resolutions, poor depth perception, complexity, and the possibility of operator disorientation because of "unnatural" spatial relationships between the TV cameras and the manipulators.

TV can boast some notable successes in teleoperators, such as the Minotaur, Mobot, MRMU, RUM, and the manipulators in the North

American SETF (SNAP Environmental Test Facility). In the last instance, manipulator operators prefer TV viewing to direct viewing through a hot-cell window (Henoch, 1964). While TV may be more convenient than direct viewing in many operations, everyone agrees that direct vision aided by purely optical devices (periscopes, telescopes, etc.) yields the sharpest, most realistic images.

Conventional 2D, black-and-white TV gives a rather limited representation of the complex scene a manipulator operator wishes to interpret. 3D, color TV was tried in the early 1950's at the AEC's Nuclear Reactor Test Site (NRTS), in Idaho, as part of the Aircraft Nuclear Propulsion (ANP) Program. The major problems encountered with 3D, color TV were: reduced light transmission, diffusion effects with the colored images, and the cumbersome camera arrangement (Morand, 1961). In the end, a black-and-white stereo TV system for which the operator wore polarized glasses was adopted to give some degree of depth perception to distant scenes.

In retrospect, the ANP experiment was premature and an unnecessary setback for TV. Because of early equipment difficulties with 3D, color TV, even 2D, black-and-white TV fell into disfavor. Many years have passed since the ANP experiment; advances in the TV art would insure that the experiment would be more successful if tried today.

Recent refinements of TV have hardly been exploited at all. Only 0.5 percent of the human eye's retina is utilized by conventional TV. Too little research has been done to couple man to TV in a more comfortable and successful symbiosis.

Most teleoperator TV installations are custom-built. Nevertheless, there is a decided advantage in using commercial TV standards that specify the number of lines, number of frames per second, and so on. A large array of highly reliable commercial TV equipment has been developed. Many miniaturized cameras are available, and important accessories, such as remote pan-and-tilt units, are readily adaptable to teleoperator work.

Two major types of TV cameras exist: the image orthicon and the vidicon. The latter is lighter, cheaper, more stable, more rugged, and has longer life. The image orthicon, though, possesses greater sensitivity and resolution. For space and undersea applications, where weight and ruggedness are critical, the vidicon is the favorite choice. Vidicons have "snapped" spectacular pictures of Mars and the Moon. Other vidicons have been adapted to underseas work. Oceanographic Engineering Corp., for example, makes a vidicon camera that can be rated for a depth of 40,000 feet. This particular camera is built around a Type 7282A vidicon with a peak sensitivity of 450 m$\mu$, a wave-length at which seawater is very transparent. An automatic light compensation ratio of 10,000–1 is also provided.

In outer space, the first teleoperator applications are likely to be on orbital vehicles, such as the Space Taxi mentioned previously. Such vehicles, however, will rely mainly on windows and direct viewing. Essentially the same situation exists in the sea; the small submersibles, which have missions so similar to those of their space counterparts, are provided with adequate viewports in the neighborhood of the manipulator arms.

In its study of a teleoperator-carrying repair and maintenance satellite, General Electric proposed the system shown schematically in Fig. 6.2. The orbital task was illuminated with three 5-watt incandescent lamps

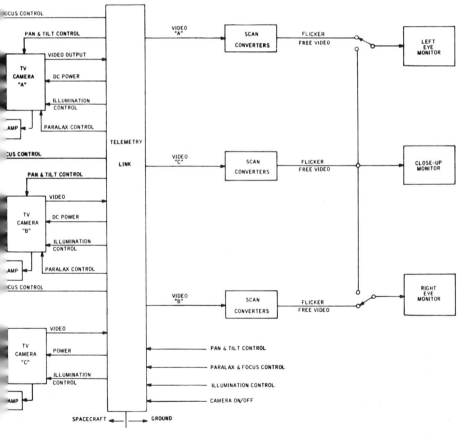

Figure 6.2 Television and lighting schematic for the General Electric repair and maintenance satellite.

with reflectors. Reflector-diffusers and automatic light control devices were added to the system. A lamp was mounted on each of the three television cameras. Two cameras on movable platforms can be focused, panned, and tilted by the operator back on Earth. Of course, the teleoperator itself can be called in for help in camera adjustment should trouble develop with the normal terrestrial controls. A third camera is attached to a semirigid tether; it can be positioned by the manipulator for close viewing of the work area.

Remotely controlled zoom lenses and pan-and-tilt mechanisms are only crude aproximations to what an operator would like. It seems wasteful to distract the operator with a second joystick or another set of console switches just for camera control. What is needed is a TV camera linked or servoed to the operator's head and/or eyes just as the manipulator hands and arms are connected mechanically, hydraulically, or electrically to the operator's hands and arms. Retaining manipulator terminology, we would call this a "master-slave" television system.*

As early as 1958, Philco Corporation engineers constructed a master-slave TV headset system with two degrees of freedom—pan and tilt (Comeau, 1961). In 1965, the Argonne National Laboratory group headed by Ray Goertz produced their Mark TV1, another master-slave with two degrees of freedom (Goertz, 1964). The success of TV1 encouraged ANL to build TV2, possessing the five degrees of freedom illustrated in fig. 6.3. Operating TV2 in conjunction with a pair of ANL Model E2 electric slave-master arms, an operator would bridge distance or a hot-cell barrier with a total of $5 + 7 + 7 = 19$ degrees of freedom, five of them associated with vision.

The TV2 television is a General Precision, Inc., GPL Precision 800 closed-circuit system using a vidicon, commercial standards, and a Zoomar f/2 lens. Both the camera and the TV monitor viewed by the operator are servoed to the operator's head. When he turns his head to the right, the camera in the operating space turns with a one-to-one correspondence and so does the vertically suspended TV screen in front of him. The whole effect is remarkably realistic. Even more realism might be achieved if a miniature TV tube were mounted directly in the operator's helmet or if a wide-screen panorama were presented.

The human head has six degrees of freedom, just like the hand of the ubiquitous Mod-8 master-slave manipulator (minus the grasp motion, obviously). In the ANL TV2, the head-cocking motion was intentionally

---

* In principle, optical equipment, such as scanning periscopes, could also be controlled by the operator in a master-slave fashion. Since there is no force feedback in these "master-slave" viewing systems they are actually "unilateral."

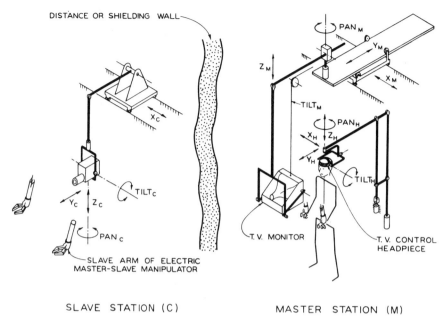

**Figure 6.3** Schematic diagram of the ANL TV2 master-slave TV control. When the operator moves his head in any of five degrees of freedom, the TV camera moves a corresponding amount.

left out because it is seldom used by an operator and presents no new view of the scene. The up-and-down and side-to-side translational motions of the whole head are particularly useful in helping the operator gain depth perception. The back-and-forth degree of freedom permits the operator to move closer to the work when desirable and vice versa. (Fig. 6.3)

The ANL TV1 camera was closely coupled to the motions of the operator's head. Experience showed this was undesirable because the slightest motion would be communicated to the camera, resulting in picture blurring. Later, on TV1 and TV2, a deadband of approximately 10° was permitted in the head's pan-and-tilt motions along with equivalent translational "play" in the other three degrees of freedom.

A radically new approach to providing the operator with both a wide field of view and high resolving power has been developed for DoD by the TRG Division of Control Data Corp. In the TRG concept the TV image in the central 8° of the field of view is eight times larger than the wide-field range image which occupies 68°. In other words, a magnified image is superimposed upon the wide field of view.

In the oculometer and the various other devices that detect eye motions we have signal sources that can give us even finer control over remote viewing equipment. It is not desirable to convert every flicker of the eye into a command signal to a TV camera, but gross motions of the eyeball might profitably be harnessed to a camera-pointing control system. This type of control would be useful if the TV camera (or some other visual sensor) had a very narrow field of view. When one already has a 30° field of view (e.g., the ANL TV2), most targets within "eyeball range" are already before the operator on the TV screen.

The teleoperators of the future that are dispatched to explore planets and the undersea by proxy, leaving man behind on Earth, will probably carry viewing systems based on the television systems control technology pioneered by Philco and Argonne National Laboratory. In particular, the ANL master-slave TV2 television system, when combined with the force feedback (feel) of electric master-slave arms and hands, is one of the most intimate examples of sensory integration we have in man-machine systems.

## ACOUSTIC SENSORS

Sound provides a sensory channel to the operator's brain that is separate and distinct from the visual and "feel" channels that predominate in most teleoperator work. Sound can thus serve well for alarm signals, activated, say, by a microphone near the manipulator hands to tell the operator that something has been dropped. The auditory sensory channel requires a much smaller bandwidth than TV.

Sound also can be used to "illuminate" a target (to use radar terminology). To illustrate, sonar gives range, range-rate, and directional information. Even better, "imaging" sonars allow the operator to "see"—crudely—in the ocean depths despite murky palls of sediments that would render visual viewing systems useless.

Several contemporary hot-cell manipulatory systems, such as Minotaur, incorporate microphones as signal pickups to warn of dropped equipment, malfunctions, collisions, etc. Such applications of sound are useful but strictly supplementary in character.

Kama, Klepser, and others have experimentally evaluated sound as a means of improving the performance of a manipulator operator (Kama, 1964, Klepser, 1966). In several experiments they employed Mod-8 mechanical master-slaves and put subjects through simple manipulatory tasks using vision supplemented by microphones. Such variations as monaural sound, stereo sound, white noise that masked all other sounds, and earplugs were tried. The general conclusion from human-factors studies like these is that the auditory channel is of little if any value to the operator, *unless* the task and microphones are specially designed to provide *significant* auditory cues.

Imaging sonar appears to hold more promise for teleoperators than simple microphones. An imaging sonar is analogous to television except that ultrasonic sound (around 500 kHz) substitutes for light. Because the wavelengths of these sound waves are so much larger than those of light, image resolution will be worse. But amidst clouds of sediment, sound will penetrate where light will not. A number of projects in this country and abroad are directed toward refining "vision with sound."

An imaging sonar might work like this: a sound transmitter (floodlight analog) would illuminate the target area and manipulator arms. Reflected sound waves would be focussed with a sound lens on an array of tiny hydrophones (perhaps small piezoelectric crystals), to form an image of the scene; the crystal array would then be scanned by an electron beam or by phase techniques (similar to phased radar arrays). The electronic signals then could be fed to a cathode-ray tube to create a visual image of the scene for the operator.

Ultrasonic sound signals are attenuated more rapidly in seawater than conventional sonar signals which have ten times the wavelength. Nevertheless, they easily penetrate ten or twenty feet. Conventional sonar could, of course, locate the target in the first place.

Sound lenses must have large diameters to focus the incoming, long-wavelength sound waves into a sharp image. At 500 kHz, an eight-inch lens is needed to provide a resolution of 1°—a very coarse image by optical standards. Sound lenses are made from plastic or a liquid, such as carbon tetrachloride, encased in a hollow plastic lens.

Before imaging sonars can be used to guide manipulator operations on the sea bottom, much more development work must be completed.

## TOUCH SENSORS

The touch sense has already been made an integral part of teleoperator technology in terms of primitive microswitches and the more sophisticated bilateral master-slaves. A bilateral teleoperator, with its force feedback, can locate a target and reconnoiter it crudely with touch. It can also evaluate the target from the kinesthetic point of view; that is, it can estimate weight and resistance to motion.* But there are other aspects of

---

* The tactual (touch) sense differs from the kinesthetic (proprioceptive) sense.

the touch or tactual sense channel that need exploring, such as shape and texture recognition. These are beyond the reach of force-feedback systems with only seven degrees of freedom.

One way to approach the sensing of shape and texture is through the addition of more bilateral degrees of freedom in the teleoperator hand. To a very limited extent, Handyman does this with its coarse "finger" articulation. A curved surface can be distinguished from a flat surface in this way. Greater articulation, of coarse, would refine the information fed back to the operator. There are practical limits, however, to the number of servoed finger joints one can put in a hand of normal size.

The next logical thought is to distribute dense "tactual" arrays of tiny force transducers over the teleoperator hands and fingers. Such an array might be modelled after the piezoelectric grip sensors that have been incorporated in the jaws of some unilateral manipulators. The totality of force signals generated by such a tactual array would depend upon the shape of the object being held and, if the transducers are small enough, upon the object's texture or roughness. In practice, the array of piezoelectric crystals could be scanned by an electron beam and the electric signals could then be rendered as a visual display or, even better, as displacements or forces on the surface of the master hand. A true bilateral master-slave hand/finger surface with hundreds of degrees of freedom is possible in principle.

Some other potential tactual transducers are air jets, tiny strain gages, and rubber fabricated with dispersed carbon. Bliss, at Stanford Research Institute, has investigated piezoelectric and air-jet approaches under NASA and Air Force contracts (Bliss, 1965).

Another class of transducers that has a potential for yielding tactual information depends upon the generation of visible effects through the deformation of a continuous rather than quantized "sensitive" surface. Moiré patterns and photoelastic effects immediately come to mind. Sheridan's group, at M.I.T., has explored several possibilities for NASA (Kappl, 1964; Strickler, 1966). The basic approach involves illuminating the deformed sensitive surface, as shown in Fig. 6.4, detecting the resulting pattern on a TV camera watching through a fiber-optic bundle, and then displaying it to the operator on a TV screen. Besides Moiré patterns and the various photoelastic effects, one might use pliant opaque "skin" with a mirror surface on its inside. By projecting a geometric pattern on the mirror surface from below, surface distortions due to the pressure of the target can be discerned as pattern distortions by the operator watching through the TV camera. The major difficulty with these "deformation" sensors is that the operator must be educated in the art of interpreting Moiré and photoelastic patterns. The relationship between the patterns

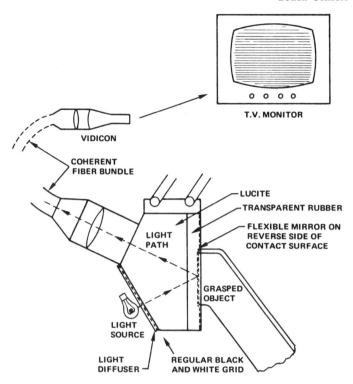

Figure 6.4 Schematic of the M.I.T. optical touch sensor.

and the shape of the grasped object is frequently subtle and quite artificial compared to the more direct bilateral feel of the object through force feedback.

One can only guess the practical effectiveness of deformation devices while they remain in the research phase. Obviously, something better than force feedback will be needed as teleoperators are applied more widely.

Man's visual and touch senses are the most useful of all his senses in projecting his natural dexterity through barriers and across distances. Considerable progress has been made in both areas in the last two decades. Examination of the other senses shows that ordinary listening does not seem to be of much use, though sound in promising as a target illuminator underseas. Taste, smell and the other less-well-defined human senses apparently have had no roles in the advancement of teleoperators. The major areas for sensory development seem to be (1) in the closer *visual* coupling of man to the working area and (2) the many-fold multiplication of the amount of force or tactual information reaching the operator. The improvement of man's "presence" or involvement in a hostile environment or at some distant place depends upon better seeing and feeling.

## DISPLAYS

In the broadest sense, a teleoperator display is the output station for all sensory information fed back to the operator. The display is the output counterpart of the input hardware described in the preceding sections. Together, controls and displays connect man to machine and vice versa; they are interface devices.

The word "display" connotes a pictorial, visual view of some scene or situation. Control engineers broaden the meaning to include abstract and symbolic displays, which represent scenes or situations in less natural terms, such as a digital distance reading or a stylized manipulator configuration. In teleoperator engineering, the concept of a display must be expanded to include the complete panorama of man's senses; past, present, and predicted future; couched in anthropomorphic or abstract language. A TV scene of the interior of a hot cell is a display; but so is the force feedback in the arm of an electric master-slave; so is a warning buzzer signaling that a joint's limit of travel has been reached.

Display design is a field of great importance in the engineering of aircraft, manned spacecraft, and submarines, where the operator must be aware of a great deal more than he can perceive looking through a window or porthole. In fact, windows and portholes are not used at all on some vehicles; instead, a "picture" of the environment is drawn by radars, sonars, and other sensors.

Teleoperators are manipulatory and sometimes pedipulatory and mobile; vision is crucial to good performance in most cases; force feedback and tactual feedback are desirable where they can be obtained at a reasonable price. The other senses, such as sound, are not nearly as important. Teleoperator displays in use today differ little from those in advanced aerospace and undersea vehicles. In fact, they owe much to the display theory and hardware developed for these vehicles (NASA, 1967).

Ideally, a teleoperator display would show the environment of interest (including the objects to be manipulated, local temperature, and other factors) and the present position or status of the teleoperator. This type of information gives the operator a seat-of-the-pants feel for the situation.* Good displays project the operator into the place where his ersatz

---

* Teleoperator technology will eventually be able to supply multiple operator feedback terminals so that many scientists could participate in, say, remote lunar exploration; although only one person would be the true operator, of course.

hands and legs are working—he *identifies* with the situation. We have stressed in earlier chapters that manipulation also requires planning and strategy formulation. We therefore must make room in our definition of displays for portraying *executive* information that will help the operator make decisions. Two special types of displays that fall in this category are the predictor and historical displays; one looks into the future using known physical laws; the other searches the past for relevant information. Teleoperator displays can and probably will be much more generalized than the hot-cell windows and closed-circuit TV typical of most extant teleoperator applications.

Display engineering has been largely intuitive in the past. In the case of teleoperators, the basic tenet has been to make the display as real as possible; that is, to duplicate the sight, sound, and feel of the task as faithfully as possible. This philosophy is a natural corollary to the assertion that teleoperator controls should be as anthropomorphic as possible. Both of these views are being challenged today.

While no formal teleoperator display theory exists, some progress has been made recently in formalizing display theory for use in conventional manual control situations; i.e., aircraft and undersea craft. Kelley's book (Kelley, 1968) and a recent paper by McRuer and Jex (McRuer, 1967), are representative of this work. Most display theory deals with forced-input tracking situations and offers little to the designer of a teleoperator display.

Conventional display theory does offer a checklist of points to consider and pitfalls to avoid in teleoperator display design:

1. Noise seriously degrades displays. A reasonable signal-to-noise ratio must be obtained in all sensory dimensions.
2. The effectiveness of a display is reduced by intermittence; that is, the reduction in time intervals when the display is active or sensed by the operator (Newell, 1959). This factor applies to the time-multiplexing of display information and the sampled-data aspects of the operator as he shifts his attention from one display to another.
3. Time-delayed feedback is highly disruptive as mentioned in Chapter 4. Predictor displays may minimize this effect.
4. Visual display parameters of magnification, framing, color, dimensionality, contrast, brightness, etc. must be considered (Smith, 1966), although few objective data are available to guide the designer.

Once it is admitted that natural, pictorial displays convey only part of the information an operator desires, the way is open to symbolic displays. The word "symbolic" is used here to mean non-pictorial. A simple warning light indicating that a manipulator limit of motion has been reached for a manipulator is a symbolic display because an "on" light is a code

**164    The Sensor Subsystem**

signal understandable to the operator—a signal conveying far more than one bit of information.

The basic function of a display is to provide the operator with enough information to make decisions; this information need not be pictorial to be useful. In fact, manipulation can be accomplished without natural, pictorial feedback at all. Computer-controlled manipulators never "see" their targets at all. Conceivably, a human operator could manipulate objects given enough force and tactual feedback plus a good repertoire of executive signals, although performance might suffer considerably without vision.

Besides warning lights and other status signals, what other kinds of symbolic visual displays might be useful in teleoperator work? Perhaps the most obvious type would be an abstract portrayal of the working environment, its targets, and the teleoperator arms and hands—a substitute for a natural view, which might be unobtainable. The scene could be drawn on a cathode ray oscilloscope tube (CRT) in stylized fashion, showing the manipulator and its targets vividly in three dimensions, possibly color-coded for easy identification, noise could be suppressed, and target data could be inserted verbally near the target image on the CRT (the air-traffic-control example again) (Fig. 6.5). The Computer Image Corporation has been pioneering this type display. Such an abstract,

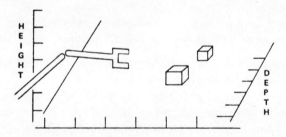

**Figure 6.5**  A possible abstract visual display indicating the configuration of a teleoperator and the external environment. Such a display need not be drawn from visual data alone; that is, sonar, radar, and status sensors can be employed.

coded representation might be much easier to work with than a natural view of the scene. Furthermore, this kind of display could be drawn from many different kinds of sensory inputs: iconoscope, radar, sonar, and, of course, status data. There would certainly be anthropomorphic aspects to an abstract display in terms of spatial correspondence, but we have no experimental assurance that anthropomorphism is required.

Few symbolic or abstract teleoperator displays have been built yet, so we continue primarily in a prophetic vein, buttressed by some anticipatory work done here and there for other applications.

Symbolic displays are part and parcel of everyday life; viz., fuel gauges and speedometers in automobiles. A symbolic approach to manipulation is not hard to imagine, though there is no proof that it would be effective. Most manipulatory tasks can be described in terms of seven dimensions; we might build a symbolic display along these lines. The three degrees of freedom representing the position of the hand relative to the target could be portrayed on a CRT-drawn set of Cartesian axes. Similarly, the hand orientation—three more degrees of freedom—could be displayed as a vector relative to the orientation of the object. Finally, hand closure around the target, the seventh degree of freedom, could be represented by a vise-like sketch. The grip in pounds could be displayed numerically next to the grip display.

Why would one want to employ symbolic or abstract displays instead of honest, natural pictorial dsplays?

1. The natural visual display may possess noise, distortion, and bad contrast. (Signal processing can clean it up to some degree.)

2. A natural visual display requires an immense quantity of information—a large bandwidth. On a lunar spacecraft, for example, signal processing equipment can eliminate all data in the natural scene except those pertaining to the targets and the teleoperator configuration.

3. In some instances, there is no natural visual display because natural and artificial lighting are absent.

4. Symbolic and abstract displays may lead to better performance of teleoperators. (A contentious statement.)

An extreme example of abstract, symbolic displays is the teletypewriter employed in supervisory control. An output device as well as an input device, the teletypewriter can print out manipulator configuration coordinates, the geometric relationship of the targets, and status data—in fact, anything we wish to know about the teleoperator and its task. Of course, there is no anthropomorphism in the printout of a teletypewriter; it is hard to imagine how an operator might identify himself with the task. Operation would certainly not be natural in the sense of everyday experience. Some people, however, identify well with symbols and mathematical relationships. A matrix is as real to them as an actual force on the target. Manipulation in this case would be much like playing chess without a chessboard—some people can do it.

The hardware available and under development for the display of abstract and symbolic information, like the television systems used for natural visual displays, is beyond the scope of the survey. The variety of media is large and gives the teleoperator designer ample opportunity to explore new modes of machine-to-man communication. We list some types of visual displays:

| | |
|---|---|
| Thermochromic | Photochromic |
| Fluidic | Magnetic |
| Electrostatic | Laser |
| Plasma | Electroluminescent |

A NASA report surveys the state of the art for these types (NASA, 1968). The more conventional CRT and projected large-screen displays are discussed in Poole (1966) and Auerbach Corp. (1968).

## Visual Predictor Displays

Displays which help the operator predict the future are helpful in dynamic situations, such as high-speed piloted aircraft and submarines. Manipulators ordinarily move so slowly that predictor displays are of little importance. The major exception occurs where significant time delay exists. (See Chapter 5 for the effects of time delay.) In cislunar space, on the Moon, and beyond, teleoperator performance can be enhanced if the operator back on Earth has some sort of predictor display that estimates the consequences of his actions before he issues commands to the Earth-based transmitter.

Predictor instruments look ahead in time by constructing "models" of the situation—primarily models of the machine and its environment. The model, possibly an electrical analog, is then run faster than real time (that is, ahead of real time) and its performance is displayed for the operator. In many human tasks, a person performs these computations in his head intuitively. In guiding an automobile around a curve, the driver projects his vehicle's position as a function of time for various combinations of control actions.

The model of the situation employed by a predictor instrument is usually displayed visually. However, there is no reason why force and tactual feedback cannot be predicted for the operator. In fact, if predicted force feedback could be added to predicted visual feedback on the same time scale, the operator would have excellent grounds for decision-making.

A visual predictor display must be abstract or symbolic because there is no knowledge of the real natural world of the future—only projections. It is customary, however, to display projections in time in anthropomorphic fashion, say, as a projected vehicle track on the actual televised scene.

Aircraft instrument panels have long utilized time-derivative (rate of change) data in helping the pilot maneuver his craft.* Ziebolz and Paynter discussed the possibility of improving upon simple derivative information by employing fast models or analogs of the entire system (Ziebolz, 1953). In the early 1960s, Kelley and his associates developed a Predictor Instrument for the Navy to help control submarines (Kelley, 1962). These ideas form the basis for teleoperator predictor displays.

Because of its historical importance, we sketch a few details of Kelley's Predictor Instrument. Fig. 6.6 shows the block diagram for this device.

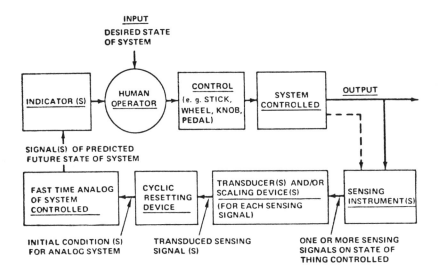

Figure 6.6 General block diagram of the Predictor Instrument. (Courtesy of C. R. Kelley, Dunlap and Associates.)

The heart of the Predictor Instrument is a miniature computer (an analog computer in this case) that models the system. As the sensors feed in information about the present, it predicts the future for various "degrees of freedom". The operator "sees" the future as a function of time and takes whatever action seems appropriate. The original purpose of the Predictor Instrument was not to overcome signal transmission time delay but rather to offset the operator's reaction time and warn him of future consequences that he might not anticipate from real-time data alone.

* The use of derivative information in generating displays is termed "quickening" and the quickened display is often called "augmented."

NASA and the Air Force have investigated predictor displays for use in orbital rendezvous, an operation where terrestrial vehicle experience is not too helpful (Kelley, 1964; McCoy, 1965, 1966). These studies employ fast-time models for prediction; again the objective is to help the operator in a complex real-time situation. Air Force-sponsored simulator studies confirm that a predictor display materially helps the astronaut.

More germane to the teleoperator time-delay problem in outer-space are the studies of predictor displays for lunar vehicles. Dunlap and Associates, Stanford University, General Motors Corp., and others have completed studies and simulation experiments (Arnold, 1963; Miller, 1966). Again, a fast-time model of the physical system constitutes the basis for prediction.

The only teleoperators to operate under time-delay restrictions to date have been the Surveyor surface samplers (see Chapter 5). Operations with the surface sampler were slow and deliberate and made use of the move-and-wait strategy. The primary display aiding surface-sampler manipulation of the lunar soil was pictorial, using the pictures taken by the Surveyor vidicon camera. Because of the 1.3-second signal-propagation time delay and the time required to scan and transmit the vidicon image, the display was what we might call "historical" in nature. Each picture was several seconds old at best. The operator of the sampler could, of course, examine as many of these still photos as he wished, but they gave him little identification with the dynamics of the experiments.

Movies made after the mission from successive Surveyor pictures have added real-time dynamic insight to the sampler operations. By showing several hour's pictures in a few seconds, the motion of the sampler and soil movement can be seen. In effect, the human brain melds the time-separated photos into a smooth whole. In future lunar operations, sped-up historical displays may quickly recapitulate the last hour of motion to lend reality to the present scene. In planning his next move, the operator could command this review of past operations at will; his brain could then project consequences of his actions better. Time delay is not eliminated, of course, but time seems compressed to terrestrial scale and the operator can use his worldly experience to predict what might happen for each prospective command.

### Force Feedback and Tactual Displays

Next to vision, force feedback to the hands, arms, or legs of a teleoperator is the most important type of "displayed" information. The mechanical and electromechanical bases for force feedback were sketched earlier. Cables and servo motors force slave arms to follow master arms and vice versa when the slave arm encounters an object. The "display," of

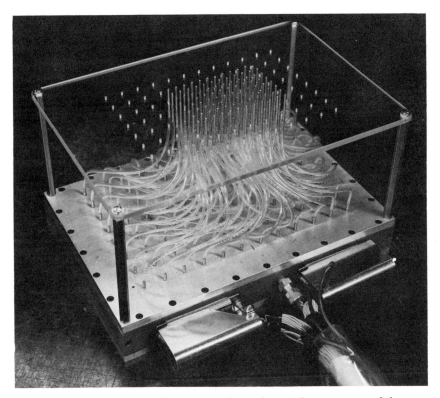

**Figure 6.7** A 12 × 8 array of air-jet tactual stimulators. The active area of the array is finger-tip size. (Courtesy J. C. Bliss, Stanford Research Institute.)

course, is the totality of forces and pressures applied to the hands, arms, and legs of the operator.

In tactual (or tactile) feedback, the situation is complicated by the fact that a well-defined, two-dimensional field of pressure stimuli is desired. Touch feedback devices thus take on some of the features of visual displays. As mentioned earlier, some tactual displays are actually visual in character. Here, we deal only with those displays that stimulate the surfaces of the fingers, although it might be argued that vibrator alarms, such as those associated with artificial limbs (see Chap. 5) might also be considered useful tactual displays for teleoperators.

Bliss and his colleagues at Stanford Research Institute (S.R.I.) have developed tactual displays for a wide variety of potential applications (Bliss, 1966, 1967). Bliss' group has constructed reading machines for the blind which convert printed letters into tactual displays that can be read by the fingers. If the fingers can discern the shapes of the letters from a tactual display, the same displays could give an operator a good sense of feel in remote manipulation.

Bliss and his associates have worked with electromechanical, piezoelectric, electrical, and air-jet stimulators. The air-jet stimulators have proven successful and have been employed in many S.R.I. experiments (Fig. 6.7). Airjet stimulators arrayed 12 × 12 have been built finger-tip size—this is the array that resolves letters of the alphabet. One can conceive of such arrays being built into the master hand controls of advanced master-slaves, with each of the 144 stimulators actuated by a corresponding pressure-sensitive spot (perhaps a piezoelectric crystal) on the slave hand.

How useful would tactual displays be if visual displays and force feedback were already applied to a given problem? Intuitively, one would say tactual feedback must be beneficial; but no one knows for sure. Bliss' human factors studies with tactual arrays have indicated that the human delay time with tactual displays alone is appreciably longer than for an equivalent visual display alone. However, human reaction time when visual and tactual displays are used simultaneously is shorter than for either display alone. Some manipulatory experiments will have to be made to determine the true utility of tactual displays. Any performance advantages would have to be weighed against the increased complexity of the teleoperator system and the engineering difficulty of installing the sensors and stimulators on machine and man.

# VII
# THE ACTUATOR SUBSYSTEM

When a signal for motion is received via a teleoperator's communication subsystem, the actuator subsystem responds by applying forces or torques to the appropriate joints in its array of hands, arms, legs, and other devices. Three classes of force and torque generators are common in teleoperators:

> *Mechanical linkages:* cables, tapes, filaments, gears, drive shafts, ball screws.
> *Hydraulic and pneumatic devices:* pistons, motors, servos, McKibben muscles.
> *Electrical devices:* solenoids, motors, servos, stepping motors.

Magnetic and electrostatic forces are also available to the designer but they are relatively weak and are employed rarely (Desroche, 1961).

There are two parts to a teleoperator actuator; these are the force/torque generator and the "switch" that receives the command from the operator and applies power to the force/torque generator. The actuating signal may be mechanical, hydraulic, or electrical, depending largely upon the specific application. In principle, actuators can be electrohydraulic, all-electrical, all-hydraulic, all-mechanical, or almost any combination of signal type and force/torque generator. Some actuators, of course, are more suited to some tasks than others. Table 7.1 shows the six combinations emphasized in teleoperator design.

A manipulator is bilateral if force and motion can be transmitted both ways to some degree, that is, from operator controls to actuators and vice versa. If one moves the slave arm of a bilateral master-slave, the master arm should also move. By this definition, most all-mechanical master-slaves should be bilateral because input and output are rigidly connected. When tapes, cables, and shafts transmit the forces, even in simple, tongs, the operator can usually "feel" what is going on at the "slave" end; he can usually move the master end by applying enough

## The Actuator Subsystem

Table 7.1  Common Types of Teleoperator Actuators

| Type of signal | Force/torque generators | | |
| --- | --- | --- | --- |
| | Mechanical | Hydraulic | Electrical |
| Mechanical | Tongs[a]<br>Prostheses<br>Mechanical<br>master-slaves | Pneumatic<br>prostheses | Electrical<br>prostheses |
| Hydraulic | | Hydraulic<br>master-slaves<br>Forging manip-<br>ulators | |
| Electrical | | Undersea uni-<br>lateral manip-<br>ulators<br>Walking<br>machines<br>Prostheses<br>Exoskeletons | Electrical<br>master-slaves<br>Unilateral<br>manipulators<br>Prostheses |

[a] In all-mechanical actuators, the operator usually transmits signals and power at the same time.

force to the "slave" end. If there is a great deal of friction, or a significant mechanical advantage between "master" and "slave," the manipulator will be *less* bilateral. With some geared systems, bilateralness in effect disappears. In fact, the incorporation of ratchet mechanisms can make an all-mechanical teleoperator truly unilateral. The same observations apply to *some* all-hydraulic and *some* all-electrical teleoperators. In the Argonne National Laboratory servoed electrical master-slaves, a force on the slave arm generates a signal that results in a force at the master arm. This two-way commerce cannot occur, though, when switch-operated motors drive manipulator joints, because a simple motor cannot generate a signal at the slave end, relay it to the master end, and create a force there. Therefore, an electrical, motor-driven manipulator is usually unilateral. However, force feedback and bilateralness may be incorporated using transducers other than the primary drive motors.

The teleoperator actuator "family tree" is portrayed in Fig. 7.1. The first branching occurs when teleoperators are classified by the type of force/torque generator used; the second florescence depends on the adjectives "unilateral" and "bilateral," while the third branching is functionally dependent (hands, feet, etc.). The sections that follow explore this "tree" and are organized in much the same way.

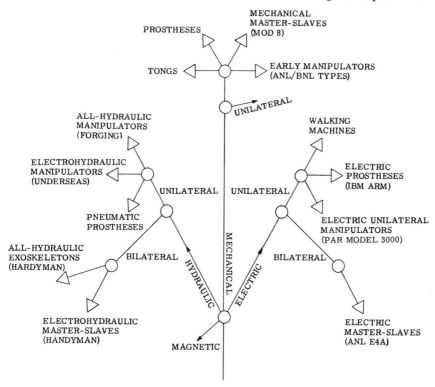

**Figure 7.1** The actuator subsystem "family tree," showing "branchings" by power source and the unilateral-bilateral attribute.

## ACTUATOR DESIGN PRINCIPLES

The actuator or "effector" subsystem mimics man's arms and hands. The simple tongs used in the nuclear industry are crude caricatures of human arms, but more advanced arms under development, such as the Serpentuator and other many-jointed arms, are even more articulated than human limbs. The actuator subsystem may incorporate some motions, such as wrist extension, that biology neglected to invent. And, of course, machines can be made bigger, stronger, faster, and more precise than men.

The actuator subsystem consists of one or more arm-hand combinations. The function of the "arm" is the translation of the hand to a desired point in space and the orientation of that hand into the desired planar position. The hand should be able to duplicate some, but not necessarily all, motions of the human hand. The most obvious function of the human hand is grasping, but anyone who has watched a manipulator operator working in front of a hot-cell window knows that hitting, poking, and pushing are as much part of the performance as picking things up.

Most manipulator discussions begin with the assertion that a manipulator arm-hand combination must possess at least seven degrees of freedom to fulfill the three basic functions of:

1. Hand translation to an arbitrary point within the working volume,
2. Hand orientation to an arbitrary plane, and
3. The grasping motion.

The first two of these functions require three degrees of freedom apiece, and grasping adds a seventh. Nevertheless, many manipulators do rather well at special tasks with less than seven degrees of freedom. Ball-joint tongs, for example, can handle many jobs with only five degrees of freedom, having sacrificed two degrees of freedom by restricting hand orientation. If an obstacle lies between the teleoperator and the target, however, more than seven degrees of freedom may be needed to reach around the obstacle and properly orient the hand. Despite these exceptions, most of the teleoperators in service today have seven degrees of freedom and the trend is toward more degrees of freedom in space and undersea applications.

How may an arm be fashioned to meet its two basic functions of hand translation and hand orientation? The human arm is an intricate series of "links" joined end-to-end by joints that can pivot and rotate various amounts. The movable joint, then, is one of the keys to articulation. A simple pivot, hinge, or sliding joint constitutes one degree of freedom. A joint can be given two degrees of freedom by adding rotation or a second pivot. A ball-in-socket joint can even provide three degrees of freedom—two angular and one of rotation. The manipulatory capabilities of the human arm depend entirely upon such a series of links (bones) and joints. Conceivably, all teleoperator arms could be built in this anthropomorphic fashion.

But why limit machines to nature's constructions? No need to, of course. Many manipulators have sliding or telescoping joints, such as the common wrist-extension feature. There are no design restrictions upon the total number of joint-link combinations in a manipulator series, or in the ways in which they are connected, or even in the number of links that terminate (or originate) at a given joint. The human wrist is really a single joint with six attached links (five fingers and the forearm). A teleoperator hand or arm may employ any number of links to suit the task at hand—always limited, of course, by cost, weight, and complexity.

In spite of the abundance of diverse possibilities for arm construction (Fig. 7.2), only a few common types have emerged.

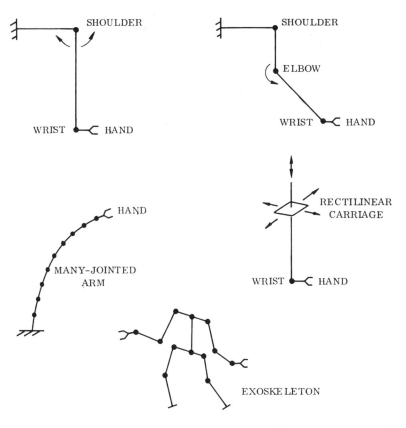

Figure 7.2 Some possible actuator geometries. Many of these geometries are illustrated in hardware form later in this chapter.

Manipulator hands are in an even more primitive state than arms. Those introduced in the nuclear field in the 1940's had vise-type hands, in which two opposing flat surfaces are brought together on the target. Except for minor changes in jaw configuration and the occasional addition of special surfaces, most of the manipulators in use today have similar hands. Beakers, fuel elements, and radar knobs are all manipulated by two opposing flat surfaces. There is little "matching" between the hand and the object.

Other hand possibilities exist in profusion. The three-jaw "chuck" hand has been proposed. The versatile hook-like hand is common in prosthetics. But so long as the designer is restricted to one degree of freedom for the hand, little sophistication can result. After all, the human hand possesses dozens of degrees of freedom. Once articulated fingers replace the vise and chuck, the hand can begin to handle round objects with finesse and generally conform to the shape of the target.

If the teleoperator hand is defined as that part which picks things up and manipulates them, there are other (perhaps better) ways than merely squeezing (vice action) or wrapping and squeezing (hand-grasp action). Pneumatic suction forces, magnetic forces, and adhesive pads made of interlocking fibers are also possible and are more common in industry than in teleoperator design.

The most common kind of mechanical linkage between the operator controls and the actuator subsystem is the flexible metal tape or cable prevalent in mechanical master-slaves. With pulleys a "pull" transmitted along a tape or cable is easily transformed into linear motion or rotation of the mechanical arm. On many manipulators where loads are heavy, mechanical motion is transmitted via link chains and gears that eliminate cable and tape stretching. When a rigid rod is substituted for the flexible cable, as it is in simple tongs, rotation or torque may be conveyed directly by the same rod that transmits linear forces. Changing the direction of a force conveyed by a rigid member may be somewhat more complex than it is with a flexible cable, but various linkages employing rigid members are available, viz., the typical vise-type hand shown in Fig. 7.3. Gears are the natural mechanisms for changing the direction of torque and rotary motion. The differential gear assembly used in one of the Brookhaven National Laboratory tongs is a good example of this approach.

Rotation and linear motion in rigid members are easily interchanged through the use of worm gears and rack-and-pinion assemblies.

The simplest hydraulic (or pneumatic) actuator is the piston that transforms a command into linear motion or force. Any linear motion, of course, may be subsequently modified in magnitude and direction by the mechanical devices. "Simple" pistons become rather complex when provided with all the valves and connections required for positive, two-way, controlled action. Nevertheless, hydraulic actuators are gradually replacing electric actuators in undersea unilateral manipulators. Important advantages of hydraulic actuators are the ease with which force amplification can be achieved, and their innate ability to transmit high forces per unit volume of actuator. Almost all heavy-duty teleoperators, such as forging manipulators, employ hydraulic actuators.

Actuator Design Principles    177

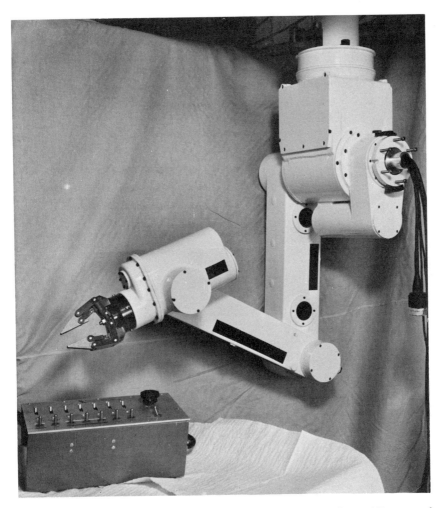

Figure 7.3  A typical switch-controlled electric unilateral manipulator. (Courtesy of R. Karinen, Programmed and Remote systems.)

An electrical analog exists for each of the hydraulic actuators; the solenoid replaces the piston, and many varieties of electric motors and servos have been developed. Electrical actuators are easy to activate and control. They are compatible with electrical signal communication and the electrical power subsystems common in teleoperator work. But electrical actuators are relatively weak. Motors, for example, must operate at high speed through long, backlash-prone gear trains to generate powerful forces. Still, the attractiveness of electrical actuation has led engineers to apply it to every field, from artificial electrical arms for the handicapped to the high-capacity wall-mounted manipulators at NRDS in Nevada.

Before dealing with specific pieces of hardware, the subject of actuator subsystem figures of merit must be broached. From the many diverse qualities used by manipulator engineers in describing their equipment, one would presume that their arms and hands possess many-sided, complex personalities; and, being extensions of man, the hardware does seem to assume a personality of its own. The list of descriptors and figures of merit that follows is at once a glossary and an intercomparison of different types of actuators. (Compare this list with the control-oriented list in Chapter 5.)

| *Descriptor or Figure of Merit* | *Definitions, Comments, and Intercomparisons* |
|---|---|
| Volume of motion | The manipulator's working volume, assuming no obstructions, is related to arm reach and its degrees of freedom. |
| Torque | Usually applied to wrist action and the ability to tighten nuts and bolts, etc., but also a property of any rotating joint. |
| Load rating | The force or lift capability available in a teleoperator arm-hand assembly. In manipulators and prostheses, the lift or load rating is that figure attainable over thousands of lift cycles. Jelatis has pointed out that at present there is no universally accepted basis for such measurements (Jelatis, 1959). Although hydraulic arms are generally used in high load situations, in principle, any type of arm can be designed for any load. The load rating usually decreases with "reach." |
| Squeeze | A "hand" rating similar to the load rating described above. The same comments apply. |
| Speed | The linear or angular rate at which a joint or the end point (hand) of a series of joints moves. Hydraulic arms are usually rather sluggish, but speed can often be |

### Actuator Design Principles 179

| Descriptor or Figure of Merit | Definitions, Comments, and Intercomparisons |
|---|---|
| | traded for force through the mechanical-advantage route. In principle, any arm can be designed to whatever speed is desired, though other factors will suffer. |
| Mass or weight | This factor is particularly important in space applications; it depends upon the load rating, working volume, and other factors. |
| Accuracy | An arm or hand is accurate if it responds to a command (say, rotate hand 30° clockwise) with some agreed-upon degree of precision. Accuracy depends upon the control subsystem to a large extent. |
| Ease of indexing | The ability of an arm-hand assembly to move into a prescribed configuration, viz., a compact, "stowed" position on a submersible. |
| Stowability | Ability to achieve a compact, flush configuration, usually within a well or compartment on a vehicle. |
| Articulateness | A measure of the number of joints and degrees of freedom. Note that too much articulateness may confuse the operator. Dexterity is usually synonymous with articulateness, though in actuality it depends heavily upon the quality of the control subsystem. |
| Stiffness | A synonym for manipulator rigidity. A stiff manipulator will tire the operator. This is an important parameter in unpowered teleoperators. |
| Inertia | A measure of the difficulty of accelerating and decelerating the actuator subsystem over and above inherent friction and the time lags caused by circuitry and linkages. Teleoperator inertia can cause overshooting and oscillations (hunting) about a target position. Too much inertia will tire the operator of a master-slave. |

| Descriptor or Figure of Merit | Definitions, Comments, and Intercomparisons |
|---|---|
| Sponginess | This is a characteristic of pneumatic teleoperators in which controls and actuators are connected by a compressible fluid. |
| Backlash | Geared force-transmission systems display this property, which is measured by the amount the actuator (or control) must be moved in the reverse direction before the commanded joint begins to move. |
| Friction | Resistance to motion over and above inertia. Friction can also tire the master-slave operator. |
| Stability | The ability to move smoothly from one configuration to another and maintain it without jitter or hunting. Depends largely on control subsystem design. |
| Sensitivity | A teleoperator is sensitive if a slight motion of the controls causes arm or hand motion. Often "play" or a "deadband" will be built into the system to prevent excessive sensitivity. |
| Drift | Electrical and hydraulic actuator subsystems may move very slightly in a continuous fashion on account of servo "leakage." |
| Cross coupling | When motion in one degree of freedom causes motion in another, cross coupling exists. This occasionally occurs in mechanically coupled systems. |
| Compliance | A measure of the match between the motion requirements of the task and the motion capabilities of the manipulator. Discussed at length in Chapter 4. |
| Maintainability/ repairability | The ease of gaining access to the actuator subsystem and effecting repairs, etc. |
| Reliability | The capability of the subsystem to operate successfully for a specific period of time. Reliability is related to complexity. The more complex electrical and electrohy- |

| Descriptor or Figure of Merit | Definitions, Comments, and Intercomparisons |
|---|---|
| | draulic bilateral master-slaves are generally less reliable than simple all-mechanical actuators. |
| Ruggedness | A hard-to-define term that usually means that a piece of equipment can survive rough treatment successfully. Strictly speaking, ruggedness is not related to load rating. |
| Fail-safe capability | When a teleoperator fails or loses power, say, in a control circuit, the actuator subsystem should maintain its configuration rather than drop objects held in the hand, etc. |
| Self-protectivity | Actuators should be designed with limit switches and other devices that prevent them from being overloaded beyond the damage point or smashing against supports, and so on. |
| Self-repair capability | Arm-hand pairs can be arranged so that one can repair the other without the necessity of men entering a hostile environment. |
| Cost | Electrical and electrohydraulic servo manipulators are considerably more expensive than mechanical master-slaves, although higher performance is claimed. |
| Power requirement | All-mechanical teleoperators require no external motive power at all, while electrical master-slaves need several kilowatts. Power is critical in space and undersea work. |
| Support-equipment requirements | Again, the electrical and electrohydraulic teleoperators are at a disadvantage because they require banks of supporting electronic gear and trained technicians. |
| Operator skill required | The effective matching of the man-machine interfaces can ease the skill requirements. |

| Descriptor or Figure of Merit | Definitions, Comments, and Intercomparisons |
|---|---|
| Resistance to the environment | Actuators must be designed to resist the corrosion, vacuum, temperature, radiation, and other aspects of the environment in which they are immersed. The simpler, all-mechanical teleoperators usually fare best in difficult environments. |
| Cosmesis | In prosthetics, particularly, the actuators should look natural and be relatively noiseless. |

This long list of actuator design factors illustrates the difficulty of teleoperator design, the multitude of tradeoffs, and the subtle interfaces. None of the factors listed above is independent of others, and there is no single over-riding figure of merit. These actuator-oriented parameters are all related in diverse, complicated, and often unknown ways to the system-wide figures of merit discussed in Chapter 4. Since no one really knows all of the interrelations, much teleoperator engineering remains intuitive and a matter of experience.

## ALL-MECHANICAL ACTUATOR SUBSYSTEMS

One of the earliest "hostile" environments that man encountered was high temperature. He quickly developed all manner of pokers and tongs for manipulating hot objects. Other "remote handlers" were constructed for working with chemicals and other hazardous materials. These are so well known that they will be bypassed here.

In the nuclear industry thought for personnel safety led first to long tongs. Nuclear radiation, though, proved impossible to attenuate sufficiently by distance alone. A way had to be found to use tongs through walls of lead bricks and concrete. The obvious solutions were to go over the barrier with jointed tongs or through the barrier with the aid of a flexible joint fixed in the wall. Both approaches met with success.

Ball-joint or ball-swivel tongs are sometimes supported in a thick ball of lead or uranium encased in steel and located in a socket in the barrier. The ball is free to rotate, although friction forces may be high. Some balls "float" on a blast of compressed air from below that reduces friction significantly. Another slightly different solution is the so-called "castle manipulator" (Ferguson, 1964). Instead of a ball, it utilizes a cylinder within a cylinder to achieve two degrees of freedom. A third degree of freedom arises when the tongs are permitted to slide back and forth through the joint; a fourth is gained when the tong shaft can rotate the hand; and a fifth, when the grasping motion is added to the hand through mechanical or hydraulic linkages. Unarticulated tongs are sometimes as long as 14 feet.

The straight, unarticulated, ball-swivel tongs can reach only those targets located within the 65° cone permitted by the joint, and then only with limited orientation of the hand. Several types of articulated tongs overcome some of these deficiencies. These tongs are usually jointed, and permit more flexibility in hot-cell operations (Stang, 1958). Models with direct spatial correspondence of motion, mirror-image correspondence, or both, are available. The driving torque for the extra joint is transmitted by means of an internal drive shaft and gearing at each joint.

The tongs just described are all "bilateral" in the sense that motion may be transmitted from either end. Interestingly enough, the mirror-image motion possible with some articulated tongs takes them out of the master-slave class, because spatial correspondence is lost, although they are still bilateral.

Through-the-wall tongs have proven very useful in the nuclear and chemical industries, but they are still restricted to relatively small operating volumes and are hampered by their lack of the full seven degrees of freedom required for dexterous tasks. Over-the-wall manipulators and additional degrees of freedom came simultaneously.

## Some Unilateral Mechanical Manipulators

Goertz has described an early over-the-wall manipulator in which most of the seven degrees of freedom were controlled by mechanical means (Goertz, 1964). This Argonne Laboratory manipulator has been termed "unilateral" because force reflection in the various degrees of freedom is attenuated to uselessness through friction and mechanical advantages. Still, in principle, force can be transmitted in both directions. This same manipulator is also "rectilinear" in the sense that the hand is positioned in two dimensions by an overhead carriage moving in $X$–$Y$ coordinates, and by a vertical column moving up and down along the $Z$ axis. Hand positioning, then, was in rectangular coordinates, and the adjective "rectilinear" became attached to all manipulators relying on overhead bridges for positioning, even though other degrees of freedom were polar.

Like Argonne, Brookhaven National Laboratory has developed several mechanical rectilinear manipulators (Stang, 1959). Models BNL–3 and BNL–4 are typical. BNL–4, for example, controlled the $X$–$Y$–$Z$ motions of the hand with cables that moved on overhead carriage. Cables attached to the operator's controls, which were full-sized analogs of the actual hand and arm, also caused rotation of the vertical column and motion of the pivoted wrist joint. The features that separated BNL-4 from the earlier Argonne mechanical master-slaves were, first, the X–Y–Z type motions that made it rectilinear and, second, the long cables and many mechanical-advantage pulleys that made it unilateral in fact, though not in theory. Although this kind of manipulator is not generally called a master-slave, the "arm" and "hand" in the hostile area mimic the motions of the controls, i.e., there is spatial correspondence. Except for the hand, the BNL-4 manipulator has few anthropomorphic characteristics. Finally, it is obvious that only the addition of electric drive motors is necessary to convert this type of manipulator into the bridge-crane electric unilateral models so common today.

### Mechanical Master-Slaves

The mechanically linked master-slaves developed by Argonne National Laboratory and General Electric under AEC auspices in the late 1940's were major advances in teleoperator technology. These master-slaves had arms and hands that looked rather like human arms and hands. The friction in the cable linkages was reduced to the point where the operator could feel what was going on in the various degrees of freedom.

The ANL Model M1 was the first manipulator built along these principles (Goertz, 1964). Replacing the ball-swivel is an over-the-wall tube suspended from a counterbalanced hinged support. The rectilinear X–Y–Z motions of the overhead movable carriage have in effect been replaced by angular and sliding motions like those seen in the ball-swivel tongs—only with the hands offset by the length of the vertical arm—and with the "swivel" now able to move along a vertical arc as the operator lifts the whole counterbalanced assembly. The three wrist degrees of freedom and the grip degree of freedom are communicated by means of cables running through the supporting overhead tube of the M1. Cable paths are short and friction low enough so that forces are reflected, and the machine is bilateral in fact as well as principle. Note that the M1 and the mechanical master-slaves covered below do not have "elbow" joints.

The biggest problem with the ANL M1 was that it was restricted to hot cells without ceilings because of the movable over-the-wall support tube. Its load rating, moreover, was only about one pound. Radioactive sources soon became so powerful that ceilings had to be put on hot cells to prevent radiation, streaming through an open top, from being reflected back down onto operating personnel. Subsequent ANL mechanical manipulators worked first through a hole in hot cell ceilings and finally through horizontal tubes high in the hot-cell walls—the present arrangement of most mechanical master-slaves in the nuclear industry. Later Argonne models, such as the Model M8, have load capacities of up to 25 pounds.

The ANL Model M8, or Mod 8, as it is often called, became the standard hot-cell manipulator in the 1950's and it still is. Commercial concerns, such as Central Research Laboratories and AMF Atomics have manufactured thousands of manipulators built around the basic ANL Mod–8 configuration.

In the Mod 8 (Fig. 7.4) a fixed horizontal tube supports both master and slave arms, which are pivoted at either end of the tube. The tube can rotate, but not slide back and forth, through a concentric tubular support built into the hot-cell wall. Up-and-down motion along the length of the arms is accomplished by tape-controlled telescope action on the slave end, a distinctly nonanthropomorphic movement. The four degrees of freedom associated with the hand are also communicated through metal tapes or cables running over a system of pulleys. Mod 8, like the M1, is bilateral in seven dimensions.

Despite the great advances inherent in the Mod–8 design, an operator can only work about one-sixth as fast with it as he can with his bare hands. Manipulator operation is tiring, too, not only because of inertia and friction at the operator's wrists but also because staring intently through a thick shielding window is a severe strain, no matter how well-trained the operator. Nevertheless, much high-radiation-level hot-cell work is being done with the help of the Mod 8 and its many close cousins.

The Mod 8 has its weak points: cables stretch, wrist-joint gears fail, and there is some cross coupling between different degrees of freedom. These problems have been overcome to some extent by commercial manufacturers. Companies such as Central Research Laboratories and AMF Atomics also have added extended-reach capability, squeeze alarms (to protect delicate objects), gas-tight seals, and other refinements. However, it is interesting that there have been no *major* changes to the basic Mod–8 design since its introduction in 1954.

The Mod 8 is really a rather complex machine. Figure 5.15a shows the CRL control hand as it appears to the operator, while Fig. 5.15b portrays the master wrist gearing that transmits the operator's applied forces to the metal tapes connected to the slave wrist.

Note that the top of the AMF master arm has counter-weights in-

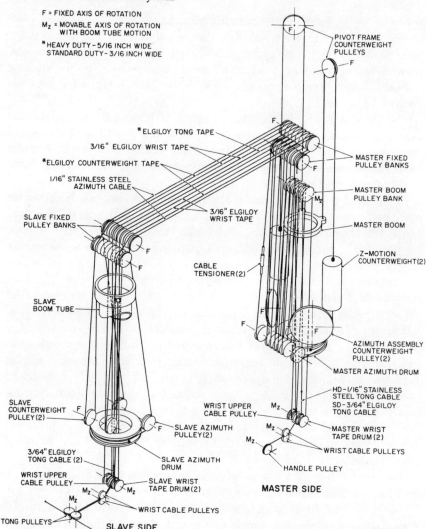

**Figure 7.4** General cabling and taping schematic of the AMF Atomics Mod-8 mechanical master slave. (Courtesy of AMF Atomics.)

stalled and that the tube piercing the hot-cell wall may contain various quantities of shielding material. In Fig. 7.5 we see the slave end of a Mod 8 with an adapter that allows it to hold a hammer. Lastly, to show the maze of cables and tapes needed to transmit operator commands in seven degrees of freedom to the slave arm and hand, Fig. 7.4 presents the general cabling and tape schematic for a Mod–8 masterslave.

All-mechanical Actuator Subsystems    187

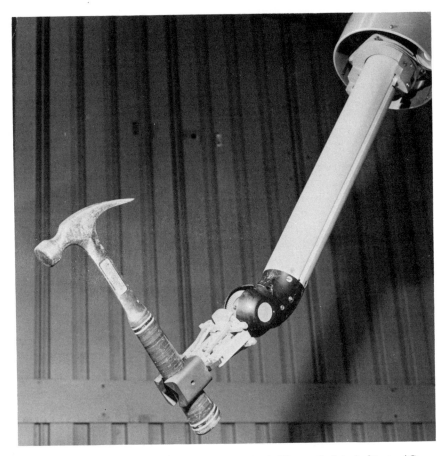

**Figure 7.5**  A Mod-8 hand with notched jaws for holding cylindrical objects. (Courtesy of J. Burton, Atomics International Division, North American Aviation, Inc.)

The Mod 8 is a workhorse of the nuclear industry, but it is not suitable for all applications. Some operators, particularly in chemical and in vacuum-chamber applications, can get along quite well with smaller, less-sophisticated master-slaves. Manipulator manufacturers have responded to this need with smaller master-slaves, such as the AMF Mini-Manip.*

The Mod 8's are mechanically connected machines, and master and slave ends cannot be separated by the distances or leak-proof barriers

---

* The AMF Mini-Manip is not a true master-slave because the Z-direction of motion is reversed between master and slave ends.

characteristic of the space and undersea application areas. The ANL electrical master-slaves, which are described later in this chapter, overcome this deficiency. Central Research Laboratories have also built gastight seals for the Mod 8, in which tape motion is converted to rotation at the sealed barrier and back into tape motion on the other side.

**Wearer-Actuated Prostheses**

For artificial limbs the criteria of design excellence are quite similar to those applied to all other teleoperators, the major exception being the property of cosmesis, i.e., looking and sounding human.

The mechanically connected prostheses introduced here are actuated by muscular action of the wearer. Because of the rigid connections from muscles to the artificial limb and vice versa, these prostheses are bilateral in the sense that an external force on the artificial limb is communicated through thong and cable to the activating muscles. Although the artificial limb is certainly anthropomorphic, a prosthesis cannot be called a master-slave device because the master end is not a physical analog of the slave end.

Distinctions among wearer-actuated artificial hands, arms, and legs depend mainly upon how much of the human body is to be replaced by the machine (Klopsteg, 1954).

Artificial hands are divided into "hands" and "hooks." Hooks are the simplest and most common of the so-called "terminal devices" (Fletcher, 1959). They are analogous to the vise-type of manipulator hand. A typical hook may show little effort at cosmesis. Some hooks are normally closed by spring action and open when actuated by the wearer; these devices have only the grip of the spring. The "voluntary-closing" hooks are also popular and are made in many sizes and shapes.

Most hooks depart from parallel vise action. For example, the Dorrance No. 5 hook opens and closes along an arc so that the open "jaws" are canted by about 20°. All hooks, as well as the hands described below, are actuated by a single cable attached to a harness worn by the amputee or to one of his muscles by a surgical process called "cineplasty." Closing forces are only three or four pounds on the average.

Designers of artificial hands (as opposed to hooks) have tried to humanize the machine. Hand engineering is still restricted by the availability of only a single control cable. This pull force must be transplanted into a hand-closing action that not only looks natural but helps the wearer do something useful, such as feeding himself. Originally, prosthetics engineers believed that curved fingers and thumb, closing in a fist-like action, would be the most useful. Experience soon proved that most manipulation is done with "palmar prehension" using only slightly curved fingers and an almost straight thumb, as in handling table utensils (Murphy, 1964). The engineering problem thus became one of moving fingers and thumb into this configuration with a single cable. Hundreds of attempts have been made to render the human hand in machine form for the benefit of amputees, using an amazing array of ingenious linkages that create varying coordinated grasping actions of fingers and thumb. The APRL No. 4 hand, designed by the Army Prosthetics Research Laboratory, is representative of these efforts. The APRL hand includes a cam-quadrant clutch, automatic locking, and three-jaw-chuck prehension (thumb and first two fingers). Other hands actuate all four fingers and the thumb, too; some boast articulated fingers. However, in prosthesis design, as in most engineering, simplicity is a powerful advantage.

A problem common to artificial hands and manipulator hands is force multiplication. An operator (or wearer) may wish to exert more squeeze than the normal actuating mechanism permits. Engineers introduced force multipliers that give the operator a mechanical advantage whenever an object is encountered by the closing hand. Added force is purchased at the price of greater displacement of the control cable.

The main function of an artificial arm is identical to that of a manipulator arm: To move the hand to the desired position in space and orient it. Unfortunately, the wearer of an artificial arm cannot bring into play the many control cables typical of the hot-cell manipulator. About all he has at his disposal is shoulder shrug, shoulder elevation, residual motion of the arm stump, and perhaps muscles brought into play by cineplasty. Although these motions can be used to power an artificial arm, the wearer cannot force his prosthesis to approach the proficiency of the normal human arm or even a master-slave.

Walking, too, is a human function amenable to mechanization with artificial limbs. The wearer of an artificial leg, however, usually does not manipulate his man-made leg save for moving his stump during the walking process. The artificial limb "steps off" and "swings" through a sequence of motions similar to those of the natural leg without any actuating cables whatever. True, a control cable may be employed by the wearer to lock the knee joint, but the amputee does not ordinarily manipulate anything.* Reluctantly, we have to exclude artificial legs and all their ingenious mechanisms from that class of teleoperators called "walking machines," but the manipulator and prosthesis industries have much to learn from each other.

---

* Some proposed artificial legs store energy (say, as pressurized gas) gathered in one phase of the walking cycle and then release it during another, viz., in "step-off."

## HYDRAULIC TELEOPERATORS

Wherever the operator cannot actuate a teleoperator directly by cables, tapes, or rigid linkages, hydraulic or electrical actuators are substituted to convert command signals into the desired forces and motions. Table 7.2 summarizes the various types of hydraulic teleoperators.

Table 7.2 Characteristics of Hydraulic and Pneumatic Teleoperators

| Type of teleoperator | Signal | Unilateral? | Bilateral? | Master-slave? |
|---|---|---|---|---|
| Pneumatic prostheses | Mechanical, electric | Yes | No | No |
| Forging and heavy duty industrial manipulators | Hydraulic, mechanical | Yes | No | No |
| Hydraulic master-slaves (Hydroman) | Hydraulic | No | Yes | Yes |
| Electrohydraulic undersea manipulators | Electric | Yes | No | No |
| Electrohydraulic master-slaves (Handyman) | Electric | No | Yes | Yes |
| Exoskeleton man-amplifiers (Hardiman) | Electric | No | Yes | Yes |
| Walking machines | Electric | No | Yes | Yes |

Hydraulic actuators are comparatively powerful per unit weight and amenable to easy force multiplication. They have been made reliable after many decades of industrial use. But, they are also leaky and prone to drift. Pneumatically actuated teleoperators are apt to be spongy; their hydraulic cousins are often sluggish. For application where strength and compactness are assets, as in walking machines and exoskeletons, however, hydraulic actuators have no peers.

### Pneumatic Prostheses

Engineers have tried to adapt external power sources to artificial limbs because wearer-actuated prostheses are weak, usually uncomfortable, and require a great deal of energy. The most popular power source is a small steel capsule filled with liquid carbon dioxide. High-pressure gas from a reducing valve develops power in an artificial limb when admitted to a piston, bellows, braided expanding sheath (McKibben muscle), or a diaphragm actuator. All of these actuators generate linear forces and displacements, simulating to some degree the action of a real muscle. Torsional devices that convert gas pressure into rotation are also available (U.S. Government, 1966). Activation of any of these actuators may be through a manual valve or an electric switch that trips a solenoid-operated valve. Control of a pneumatic arm can be made less obvious with the use of capacitance touch-switches or photocell switches. Although pneumatic arms are spongy or "soft" and difficult to control precisely, these defects may be eliminated to some degree by going to higher pressures.

The liquid $CO_2$ capsule is convenient but it does not usually store sufficient energy to enable an amputee to walk with an artificial leg. So far, pneumatic prostheses have been confined to upper extremities. Pneumatic power has also been applied to orthotic devices.

Hydraulically actuated arms and legs are possible, but they require the amputee to carry a power source, a pump, and all the requisite plumbing around with him. Nevertheless, electrohydraulic arms using water and employing hydraulic servo motors have been successfully constructed (Anon., 1965).

## Heavy Duty Manipulators

The heavy-duty manipulators employed in metal-treating plants and other operations where heavy, hot objects must be handled with a modicum of dexterity are similar to the pneumatic and hydraulic artificial arms just described. Hydraulic actuation is used in missile loaders, bulldozers, forklifts, and other heavy industrial handling equipment. But the great majority of these aids are not members of the teleoperator family because their manipulatory capabilities are far below those of a human being (Howell, 1954).

At least two small submersibles have carried all-hydraulic manipulators controlled directly by manually operated valves. These were the *Recoverer I* and the *Diving Saucer SP–300*. Later submersibles almost invariably have relied on electrical and electrohydraulic manipulators which do not compromise hull integrity with large hydraulic line penetrations.

## Hydraulic Master-Slaves

All-hydraulic, bilateral master-slaves with several degrees of freedom are rather rare animals in the world of teleoperators. Single degrees of freedom using hydraulic actuation are much more common, particularly when a strong gripping force is wanted with tongs or other mechanical manipulators. Because friction can be made low and master and slave pistons have approximately the same areas (no mechanical advantage), force and motion are transmitted in both directions; thus, the device is truly bilateral.*

The Hydroman, built by Oak Ridge National Laboratory, represents one of the few attempts to construct an all-hydraulic teleoperator. Hydroman was built for through-the-wall hot-cell operations involving heavy loads. Hydroman was given an elbow but no up-and-down telescoping action. The forearm delivered 1000 in.-lb of torque from an internal, reversible hydraulic motor. The wrist joint was a hydraulic cylinder with a rack and gear assembly to convert linear motion into rotary motion. Force reflection or feel is not transmitted back through the power loop, but through a differential feedback cylinder and a feedback force-ratio bar. Thus, Hydroman can be classified as bilateral. Hydroman is not a true master-slave because there is no spatial correspondence, but natural motions of the operator's arm and hand are communicated to analogous manipulator components through the hydraulic linkages.

### Electrohydraulic Undersea Manipulators

The combination of electrical command signals and hydraulic actuation is logical for small submersible manipulators. Hydraulic actuators perform well in high-pressure seawater and can be assigned heavy tasks. Seawater itself has been used as the hydraulic fluid for some devices such as the NEL (Navy Electronics Laboratory) manipulator. As technical interest and research and development money have flowed increasingly into undersea work, more and more innovations in the teleoperator art have come from this area.

Early undersea manipulators were either electrical unilateral machines (on *Alvin I*, the *Trieste*, and the RUM bottom crawler) or all-hydraulic (on the *Discoverer I* and *Diving Saucer SP–300*). The electrical manipulators were modified General Mills Model-150 terrestrial machines. These worked, but proved "delicate" and rather vulnerable to the deep-sea environment. Excellent results were obtained with the all-hydraulic manipulators in shallow water. Their large hull penetrations, however, would be risky at great depths. Many new submersibles use electrohydraulic teleoperators.

Hunley and Houck, in their 1965 review of underwater manipulator technology, noted that:

1. Two manipulator arms are desirable in underseas work because of the complex tasks.

---

*The GE Hardiman is an all-hydraulic exoskeleton. It is described later in this chapter.

2. In working manipulators (as opposed to specimen-collecting types), many degrees of freedom are desirable, especially wrist extension.

3. Provision for emergency jettison of the manipulator is desirable (a feature militating against all-hydraulic systems), and the jettison mechanism should not be such that manipulators may be dropped inadvertently or lost if they are not stowed properly. (Manipulators were lost at sea in early development work.)

4. Some way to confirm proper manipulator stowage is desirable.

5. Hard stops and/or limit switches of some kind are needed to prevent structural damage, even if there are slip clutches and pressure-relief valves in the system.

6. Internal leakage must be kept low to prevent drift or "creep" of the manipulator actuators.

7. External wires and hydraulic lines must be kept to a minimum because of high drag forces during vehicle towing and the possibility of entanglement with debris around work areas.

Undersea electrohydraulic manipulators tend to be larger and more rugged than their terrestrial counterparts. Another common feature is the square or rectangular, rather than circular, cross section of the arms, a characteristic resulting from such desiderata as easy fabricability and accessibility, and the desire to enclose wires, hydraulic lines, transducers, and actuators.

Hot-cell manipulators are usually suspended from an overhead support in such a way that the operator can view the hands at roughly eye level. Undersea manipulators, in contrast, are often mounted on one side or below the operator within the submersible. The arms are projected out horizontally rather than suspended vertically. Undersea arms almost invariably have shoulders, elbows, wrists, and, predictably, hands. One degree of freedom per joint seems the rule, and wrist motion is usually more limited than that in a mechanical master-slave. In other words, the arms are well articulated in order to maneuver the hand into position but the joints have fewer degrees of freedom. The wrist often has only two rather than the more common three degrees of freedom.

The Electric Boat Division of General Dynamics has developed a "prosthetic" arm control that forces the manipulator arm to take on the same configuration as the operator's arm. Since there is configuration correspondence, one is tempted to assign such a teleoperator the designation master-slave. (Note that configuration correspondence does not insure spatial correspondence because an undersea arm is usually much larger than the control arm, meaning that linear velocities are

Table 7.3 Some Characteristics of Underwater Manipulators

| Manufacturer | Underwater manipulator name or model number | Type | Assigned submersible | Design environmental conditions | Actuator power source | Manipulator-motion control devices | Terminal devices available |
|---|---|---|---|---|---|---|---|
| Programmed and Remote Systems | 150w, 300w, 1000w through 7000w | General purpose. Designed and built to specific order. | Various | Max. depth—36,000 ft. Temp. range—50° to 250°F Pressure—18,000 psi Immersion time and operating cycle—continuous | Oil-hydraulic, water-hydraulic, electric | Switches, Position controller | Parallel-jaw hand, hook hand, tong hand, saw |
| Central Research Laboratories | CRL Model B—canal manipulator CRL Model D—special underwater manipulator | Model B—Standard, off-the-shelf shallow-water master-slave manipulator. Model D—Standard. Both are general purpose. | None | | Manual, electric | Master-slave | Tong tips |

| | | | | | | |
|---|---|---|---|---|---|---|
| General Electric | Manipulator arms for the *Aluminaut* submarine | One-of-a-kind, general purpose. | *Aluminaut* submarine | Max. depth—15,000 ft<br>Temp. range—34° to 120°F<br>Pressure—7500 psi | Oil-hydraulic | Switches |
| Litton Industries, Applied Science Div. | ASD Model-162 manipulator | General purpose, underwater environment, designed specifically for *Alvin* vehicle, not a shelf item. Manipulator now in use. | *Alvin* vehicle | Max. depth—10,000 ft<br>Temp. range—28° to 100°F<br>Pressure—4400 psi<br>Immersion time—no limit<br>Operating cycle—continuous | Electric | Toggle switches | 3-fingered claw, clamshell bucket, hook |
| Westinghouse | OFRS Model No. 2 manipulator | General purpose | *Deepstar 4000* | Max. depth—4000 ft | Oil-hydraulic | Microswitches | Hydraulic-powered 3-fingered claw |

Table 7.3 (Continued)

| | Model number | Specific | | | | |
|---|---|---|---|---|---|---|
| Westinghouse | not yet assigned | purpose: recover small objects, cutting non-metal cable, and making simple cable attachments. | Deepstar 20000 (Tentative) | Max. depth—20,000 ft. Temp. range—28° to 80°F Pressure—10,000 psi Immersion time—4320 hr Operating cycle—1 hr in 12 | Oil-hydraulic | Switches actuated by depression of bar-type controllers |
| | | | | | | None |
| General Dynamics Division, Electric Boat | HT 7150—400 series | General purpose, off-the-shelf for underwater use. Units in active use as well as in various stages of design, construction and testing. Proven in sub service. | General use | Max. depth—independent Temp. range—20° to 225°F | Oil-hydraulic | Switches, hand-operated valves |

| | | | | | |
|---|---|---|---|---|---|
| Koelsch Corporation | Model 122D-1000 (modified standard model 122C) | "Dry" environment model. Proposed for general purpose underwater use—would be built to order. | None | Max. depth—Temp. range—Pressure—3000 psi Operating cycle— | Electric for moving parts, except fingers. Hydraulic for fingers | Solid-state switches |
| Scripps Institute | Benthic Manipulator I | One-of-a-kind manipulator to perform specific tasks in an oil environment under water. Now in use. | Benthic "Hive" | Depth—200 to 20,000 ft Temp. range—32°F to 150°F Pressure—100 psi to 1000 psi Immersion time—indefinite Operating cycle—continuous Operation in acid-washed kerosene | Oil-hydraulic and electric | Switches, joystick |

Table 7.3 (*Continued*)

| | | | | | | |
|---|---|---|---|---|---|---|
| Autonetics | Marine Manipulator Model I | General purpose. Design complete and prototype tested. | *Beaver MK I* | Max. depth—20,000 ft<br>Temp. range—20° to 120°F<br>Pressure—10,000 psi<br>Immersion time—10,000 hr<br>Operation cycle—no limit | Oil-hydraulic | Toggles, push buttons and joysticks, replica arm with:<br>a. Single-speed push button<br>b. Proportional-speed push button<br>c. Position servo | Hook hand, stud gun, grinding wheel, centrifugal pump, wire brush, impact wrench, cable cutter |
| American Car and Foundry | IURC M-600 | General purpose, completed | *Trieste II* | Max. depth—10,000 ft<br>Temp. range—30° to 150°F<br>Pressure—5000 psi<br>Immersion time—unlimited | Oil-hydraulic | Individual switches | Finger type |
| International Underwater Research Corp. | IURC M-100 | General-purpose prototype under construction | General use | Max. depth—unlimited<br>Temp. range—0° to 150°F<br>Pressure—unlimited<br>Immersion time—unlimited | Oil-hydraulic | Individual switches | Straight jaw others in design |

198

not the same, even though angular velocities are.) Since hydraulic actuators (pistons) are nearly always linear in their action, a rack-and-pinion mechanism is required at the joint. These hydraulic pivots are so common that we illustrate a typical actuator arrangement that provides for two-way rotation control from within the submersible (Fig. 5.12). Many variations are possible. Other linear-to-rotary actuators are the so-called "roller-chain" and "vane" actuators (North American Aviation, 1966).

An interesting design feature under development at General Dynamics' Electric Boat Division is modularity. A modular manipulator is built up from a few basic pieces, much like a Tinkertoy construction. Electric Boat can put together 28 different manipulator arms using only six different building blocks. The arms vary in length, degrees of freedom, and load capability. The modular approach can be applied to most electrical and hydraulic manipulators and even to mechanical arms, provided that suitable gear or shaft connections can be made between modules.

The demand for reliable underwater manipulators is indicated by the number of companies working in the area and the variety of hardware produced. The extensive survey conducted by North American Aviation's Ocean Systems Division for the Navy's Deep Submergence Systems Program in 1966 brought together the data presented in Table 7.3 (condensed from: North American Aviation, 1966).

Two other undersea manipulator development efforts have unique features. One is the ten-jointed electrohydraulic arm built by Marvin Minsky's group at M.I.T. Each of the joints has a single degree of freedom and is actuated by a hydraulic piston. A position transducer parallels the piston to insure that the arm assumes the same configuration as the replica control. This arm will eventually be computer-controlled.

Another development of interest is the so-called "tensor arm," conceived by Victor Anderson, at the Marine Physical Laboratory (MPL) of the Scripps Institute of Oceanography. The basic arm consists of a series of four joints (five links), each with two degrees of freedom (Fig. 7.6). The entire arm is hydraulically actuated by nylon "tendons" strung along the exterior of the arm rather than by actuators placed at each joint. A pull on one side that is not compensated by an equal pull on the other side causes the whole arm to bend in a way similar to the muscle-tendon action in the human arm, except, of course, that the tensor arm possesses two unrestricted degrees of freedom at each joint. Sensor tendons parallel the actuator tendons and give the operator position feedback. The MPL tensor arm, also called Benthic Manipulator II, can operate directly in seawater and is intended for use in

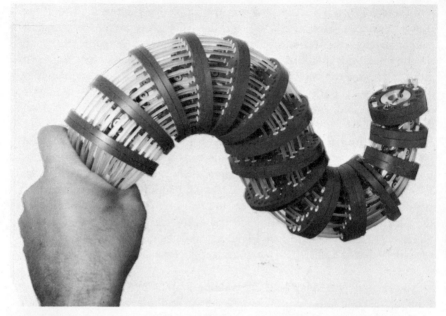

**Figure 7.6** The Scripps tensor arm. Stress on the nylon filaments actuates the arm. (Courtesy of V. C. Anderson, Scripps Institution of Oceanography.)

the MPL Benthic Laboratory "hive" for replacing electronics cards, wrapping terminals, and so on. In contrast to some of the more massive underwater manipulators just described, the tensor arm is only about 15 inches long. The novel actuation scheme is potentially very important in teleoperator design.

### Electrohydraulic Master-Slaves

The General Electric Handyman, like the first Argonne National Laboratory mechanical and electrical master-slaves, represents a milestone in teleoperator technology. Built in 1954 as part of the Air Force/AEC Aircraft Nuclear Propulsion Program (ANP), Handyman embodied a number of unique design features. Besides being the first electrohydraulic, bilateral master-slave, it was also the first to employ articulated exoskeletal master arms that conform to the operator's arms. Another "first" was the prehensile hand with built-in force reflectors. In overall dexterity and system sophistication, few teleoperators have approached Handyman, but it is a costly and complex machine.

Handyman's dexterity, of course, results from its total of ten bilateral degrees of freedom per arm-hand combination. These are: shoulder,

2; upper-arm twist, 1; elbow, 1; forearm twist, 1; wrist, 1; hand, 4. The Handyman slave arms can lift 75 pounds in their weakest position, i.e., when they are separated the farthest. Each degree of freedom in the slave arm is actuated by an electrohydraulic servo like that pictured in Fig. 7.7. Servo operation begins when a voltage increase causes the torque motor to force the reed nearer the nozzle (Mosher, 1960). Bias

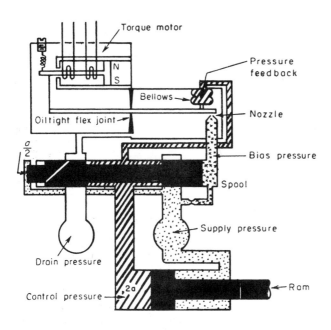

**Figure 7.7** The Handyman modified hydraulic servo with pressure feedback. The ram was packaged with the servo valve in Handyman.

pressure then increases, causing the spool to move left. Simultaneously, the control pressure is reduced because some fluid goes to the drain. The control-supply pressure differential then moves the ram to the left. At the same time, the decreased control pressure in the pressure bellows allows the reed to move back from the nozzle, thereby causing the ram force to be proportional to the torque motor voltage. Feedback signals are generated by external circuitry (as discussed in Chapter 5).

Although Handyman was never used extensively, the technology pioneered during the program has found its way into other teleoperator programs, such as General Electric's exoskeleton work (Hardiman) and walking-machine development programs.

### Exoskeleton Man-Amplifiers

Man amplifiers under study and development are either electrohydraulic or all-hydraulic master-slaves that parallel the configuration of the human body. The human operator literally works inside a two-layer mechanical suit. The inside layer consists of a master exoskeleton that follows the more critical motions of the operator whom it encloses. The heavy-duty, outside, slave layer follows the motions of the master exoskeleton that it encases. It is an onion-skin arrangement with man at the core.

Such a man amplifier sounds like a good idea, *but is it technically feasible?* Some of the earliest work on the basic concept was carried out at Cornell Aeronautical Laboratory for the Air Force in the early 1960's (Clark, 1962). These early studies concluded that:

1. Duplication of *all* human motions would be impracticable.
2. Experimentation was necessary to determine just which human motions should be duplicated.
3. Inability to counter overturning moments might limit the load-carrying capability of a man-amplifier.
4. The most difficult problems were in the areas of servo, sensor, and general mechanical design.

Further work at Cornell Aeronautical Laboratory led to the conclusion that a man could be encased in an exoskeleton with substantially fewer degrees of freedom than he possessed himself and still carry out many useful tasks without discomfort. Under a contract from the Office of Naval Research, Cornell next undertook to sketch out a preliminary design of the shoulders and arms for a man-amplifier (Mizen, 1964). This study concluded that mobility and dexterity would be adversely affected by the size of the hydraulic rotary actuators unless loads were limited to a few hundred pounds per arm.

More recently, General Electric has advanced the man amplifier concept under a contract sponsored jointly by the U.S. Navy and the U.S. Army (General Electric, 1966).* In October, 1966, General Electric concluded that, although servos are still problems, a powered exoskeleton could be constructed that would enable a man to lift 1500 pounds six feet and carry this load 25 feet in 10 seconds.

In the GE concept, the operator stands inside an anthropomorphic structure built in two halves that are joined together only at the hips by a transverse member called the "girdle" (Fig. 7.8). The exoskeleton parallels the operator everywhere save at the forearms, where the exo-

---

* Part of Project MAIS (Mechanical Aids for the Individual Soldier).

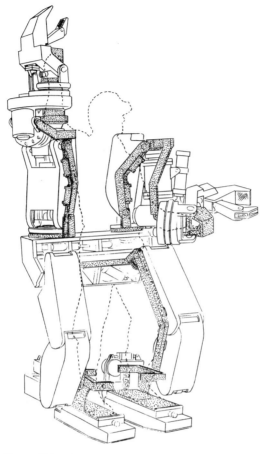

**Figure 7.8** Actuator arrangement on Hardiman I.

skeleton completely surrounds the operator, and his arms are collinear rather than parallel with the exoskeleton forearms. This forearm arrangement simplifies controls and makes it easier for the operator to identify his arm with the slave arm. The slave hand consists of one servoed degree of freedom that forces an opposed "thumb" toward a V-shaped palm-finger structure. An additional thumb-tip joint is not servoed but responds to an operator on-off switch control.

The force ratio contemplated between master and slave structures is about 25. This immediately raises a question of operator safety should the slave exoskeleton somehow run amok. In the GE design, limbs are physically linked in such a way that small master-slave errors cannot build up to do damage. Another safety feature locks all actuators should hydraulic pressure or control signals fail. Collapse of a heavy exoskeleton —carrying perhaps a 2,000-pound load—would be very hazardous without such a provision.

The articulation and dimensions of the GE man-amplifier were determined by a study of the motions that it could perform and the range of individual operators that it could accommodate without major adjustments. Operators were assumed to range from the 10th to the 90th percentile in physical size. Ultimately, the degrees of freedom and dimensions illustrated in Fig. 5.21 were selected for each side of the master-slave. With 15 joints on each side, a man amplifier could carry out most of the important human motions, save for those requiring considerable dexterity of the hand. At each joint, except numbers 10 and 12 in Fig. 5.21, hydraulic pistons were the proposed actuators. Hydraulic rotary actuators would alleviate packaging problems at joints 10 and 12. The actuator at each degree of freedom is actually a bilateral servo that reflects forces exerted on the slave members back to the corresponding master member (scaled down by 25), and then to the operator.

To compound semantic confusion, the man amplifier is a bilateral, bilateral master-slave. The first "bilateral" refers to the symmetric geometry of the teleoperator (the "bilateral" from biology) and the second to the two-way flow of motion and force between master and slave.

The electrohydraulic servos used in the exoskeleton arm system are of the bilateral (force reflecting) type. This type of servo differs from the ordinary unilateral servo in that either master or the slave may accept command signals resulting in a controlled response of the slave or master. The basic elements making up the bilateral servo used in the exoskeleton are:

—A servo valve that regulates a differential pressure, $\Delta p$, from a hydraulic pressure source in proportion to an input current, $i$.
—A hydraulic cylinder, slave "power" actuator, which converts the differential pressure, $\Delta p$, into a force that acts through the slave member to resist a load, $F_s$.
—A hydraulic cylinder, master "force feedback" actuator, which converts the differential pressure, $\Delta p$, into a force that acts through the master control to apply a force to the operator proportional to and in the direction of $F_s$.
—A control circuit, consisting of a velocity and position error transducer package "tickler," a master tachometer and the associated electronics. The electronics are required to combine and condition the transducer signals so that, when used as the control signal for

the servo valve, a stable high performance control system results (General Electric, 1969).

One of the major problems with the concept as it now stands is power consumption. General Electric estimates that the peak power consumption during normal operations would run as high as 60 horsepower. This quantity of power can be generated by a lightweight gasoline engine or a gas turbine that the man amplifier could backpack with enough fuel for several hours of operation. The weight and bulk of the power subsystem could be substantially reduced if more efficient bilateral servos could be developed. In work areas where power lines are available, man amplifiers could be "plugged in."

## Walking Machines

The man amplifiers described above are walking machines, of course, but a machine's legs need not conform to those of a man. They then may be made as large or as small as a task demands. Usually, the master-slave variety of walking machine is larger than a man.

Because of the high loads encountered with such a vehicle, hydraulic actuators predominate in designers' thinking. A walking machine, however, need not be all-hydraulic; the linkage between master and slave may also be mechanical or electrical. In an experimental balance machine, built by General Electric, the link between operator and actuators was purely hydraulic (Mosher, 1965). (See Chapter 5.)

Walking machines have been built without the master-slave relationship between operator and actuator. In walking toys and even the Space-General walking wheelchair, the operator only turns the machine on and off and steers it. In these machines, which are not teleoperators, the feet are preprogrammed to follow a specific motion, regardless of the terrain. William E. Bradley, of the Institute for Defense Analyses, has suggested substituting a computer for a human driver. A computer buttressed with suitable stored information and subroutines plus suitable sensory feedback from the feet and visual sensors could take a walking machine over unpredictable rough terrain. However, as Bradley points out, this would be "a formidable exercise in cybernetics," and much beyond the scope of this book.

Even though General Electric's dynamic balance experiment proved that a man could easily balance himself atop a servoed two-legged machine, most designers favor vehicles with at least four legs. Even a human falls occasionally and a machine without "hands" or some other aid to regaining its feet would be helpless when it fell. The big advantage of a two-legged walker would be that a single man could operate it with master-slave, force-reflecting leg controls and have his hands free for other work.

For a man to operate a quadruped master-slave vehicle, arm-control harnesses might have to be added, presumably with the man assuming a rather uncomfortable crawling position (perhaps in a slung harness). When the number of legs exceeds four (say, a hexaped), another operator working in concert with the first would be required. It might be easy for them to walk in a synchronized gait on level ground, but considerable training would be needed to enable a machine with two or more operators to traverse rough terrain. To relieve the problem of coordinating operators, some of the legs might be programmed to follow the actions of the operator-controlled legs, making the proper allowances for gait and the terrain encountered by the lead pair of legs.

Several automatic preprogrammed walking machines have been built, notably those by Shigley (chapter 2) and Space-General.* Both of these machines had legs or frames that operated in pairs on either side of the vehicle. Neither was a true teleoperator. In the Space-General machine, eight legs are preprogrammed to operate as four pairs in a sequence that keeps four legs on the ground at all times for the sake of stability. These electrically actuated automatic walkers have successfully demonstrated the feasibility of walkers, but they are far from master-slave-controlled walkers capable of traveling over unpredictable terrain.

## ELECTRICAL TELEOPERATORS

In comparison with hydraulic actuators, electrical solenoids and motors are high-speed, low-force (or torque) devices. For high strength, they have to work through long, fallible, noisy gear trains. In comparison with mechanical teleoperators, such as the Mod-8 master-slave, electrical master-slaves are more complex, more costly, and demand considerably more engineering support for maintenance, repair, etc. The simpler unilateral electrical manipulator does not have the dexterity of the master-slaves. Notwithstanding, electrical teleoperators not only survive but multiply. The reasons are many.

An amputee likes an electrically powered prosthesis because it does not require clumsy pneumatic or hydraulic hardware draped about him. Neither are there uncomfortable straps and harnesses—only simple switches, which in the case of electromyographic (EMG) control, can be activated with the flick of a muscle.

---

* Some simple "drag-lines" can be considered simple walking machines. See Liston and Mosher, for a historical discussion of walking machines.

The simplest way to pierce a barrier or overcome distance is with electrical signals. For this reason, teleoperators proposed for outer space are generally electrically actuated. Although hydraulic actuators can be made essentially leakproof, electric actuators are usually preferred in hot cells because fluids are hard to clean up from hot cell interiors and may contaminate reactor coolants, high purity atmospheres, etc.

### Electrical Unilateral Manipulators

Electrical unilateral manipulators are second only to the ANL-conceived all-mechanical master-slaves in terms of total number in use. Most are employed in the nuclear industry, though several were modified for use in the early submersibles, and some industrial applications find them advantageous (see Chapter 2). Whether the electrical unilateral manipulator arm is one foot or 50 feet long, it is basically a series of joints and links, with each joint driven by an electric motor. The operator usually actuates these points with either an array of switches or a joystick without force feedback of any kind. Sometimes proximity indicators and/or force-measuring transducers are installed at the manipulator hand, but nonetheless, the man-machine relationship is not so intimate as it is in the electrical bilateral master-slaves described in the next section.

Melton has classified electrical unilateral manipulators according to their types of mountings (Melton, 1964):

1. Overhead bridge-crane mountings.
2. Wall mountings.
3. Overhead monorail mountings.
4. Pedestal mountings.
5. Vehicle mountings.

The overhead bridge-crane mounting, with its X-Y-Z motion was employed in the late 1940's and is still very common. It was, of course, this rectilinear type of positioning that led to the common but incorrect equivalence of the terms rectilinear and unilateral. Only the three motions that position the hand at a point in space may be called rectilinear; the rotations of the hand are best termed "polar." This type of mounting is common in hot-cell work.

Wall-mounted booms, Fig. 4.1 are also rectilinear insofar as their motion along the wall is concerned; the rest of the degrees of freedom are polar. Wall-mounted manipulators are features of the immense hot cells associated with the various programs of the NASA-AEC Space Nuclear Propulsion Office, such as the nuclear rocket E-MAD building in Nevada.

Overhead monorails and pedestals are occasionally found in nuclear installations, but they are not abundant. A good many vehicle-mounted electrical unilateral manipulators, however, are assigned to emergency and routine operational tasks in nuclear facilities.

In configuration, most electrical unilateral manipulators resemble some of the all-hydraulic and electrohydraulic manipulators discussed earlier. The shoulder joint generally has two degrees of freedom (one pivotal and one rotational); the elbow joint pivots in one degree of freedom, the wrist can pivot or extend to add two more degrees of freedom, and, finally, the hand can grip and rotate, making a total of seven degrees of freedom. If the manipulator arm is mounted on a bridge-crane carriage, on a sliding column, three more degrees of freedom are added. The carriage can carry the arm over wide areas that could not be reached by the through-the-wall master-slaves seen in small hot cells. The extra mobility is purchased at the cost of the dexterity and force feedback of the mechanical master-slave. A final note on configuration: electrical unilateral manipulator arms are almost invariably mounted singly rather than in pairs—the single unit requires considerable concentration by the one operator to handle switch-box or joystick controls.

"The great utility of force feedback in assembly, repair, and maintenance work is often glossed over. Force "feel" makes the manipulator compliant to the task (as mentioned in Chapter 4) and enables two manipulator arms to work simultaneously on the same task. In contrast, two unilateral manipulator arms could not easily manipulate the same object in the absence of force feedback."

Electrical unilateral manipulators are made in all sizes and load ratings. To illustrate the general configuration, a PaR (Programmed and Remote Systems) unilateral manipulator is shown in Fig. 7.3.

The Los Alamos Minotaur—presumably so called because of its bull-like strength and man-like arms—is an exception to the statement that electrical unilateral manipulator arms are used singly (Fig. 7.9). A pair of manipulator arms plus a second pair of adjustable arms holding lights and TV cameras protrude from a sphere-like turret supported from above by a bridge-crane carriage. The Minotaur was originally built to Los Alamos specification by General Mills, Inc. A representative application is the maintenance of radioactive equipment in the shielded bay containing the Los Alamos UHTREX (Ultra High Temperature Reactor Experiment) (Wiesener, 1963).

A rather unusual electrical unilateral teleoperator is the Serpentuator (Serpentine Actuator) under development at Marshall Space Flight Center. The Serpentuator consists of links several feet long separated by joints driven by electric motors, or, in one version, electrohydraulic actuators. With maximum deflections of about 20° per joint, the teleoper-

Figure 7.9 The Los Alamos Minotaur electric unilateral manipulator system. (Courtesy of Los Alamos Scientific Laboratory.)

ator can be coiled up in circular loops 20 feet in diameter and housed in the shroud of a Saturn rocket. Using switch controls at both ends of the Serpentuator, the operators can transfer tools, retrieve objects, aid astronauts, and perform other tasks in weightless space where positive controlled motion over distances greater than a few feet are difficult. There is considerable similarity between the Serpentuator and the variable flexibility tether system designed by General Electric (Rader, 1968).

### Electric Arms

In 1945 an inventor named Samuel Alderson interested Thomas J. Watson, Sr., then president of IBM, in applying electricity to help many amputees from World War II. For about six years, using funds provided by IBM and the Veterans Administration, some remarkable pioneering work was carried out by Alderson and his coworkers (Klopsteg, 1954). Since then many individuals and organizations have advanced the art of electrical prosthetics and orthotics. Electric arms have benefited substantially from space research in terms of smaller batteries, smaller and more efficient motors, and advanced control techniques. Nevertheless they have not yet come into widespread use.

The electric arm, whether for prosthetic or orthotic applications, consists of a series of rigid links connected by motor-actuated joints. In this, there is little difference between the hot-cell manipulator and the prosthesis. The electric arm, however, must be lightweight, use little power, be quiet, and be easy to control even though its operator has no analogous limb. The electric motor is considerably more responsive, efficient, and flexible than an electric solenoid actuation of an artificial arm. In particular, the permanent-magnet, dc electric motor is lightweight and quiet. These motors are high speed (on the order of 10,000 rpm) and must be geared down before they can transmit power through a clutch to the joint. The clutches are usually of the multiple-disk friction type so that the force transmitted can be made proportional to the command signal generated by the amputee through a control cable. When the desired position has been attained, the joint must automatically lock itself.

Since two-way joint action is required and the amputee's signaling muscle usually produces only a unidirectional signal, the control logic must be such that a series of shoulder shrugs, for example, will be properly interpreted as go, stop, and reverse signals. Since control-signal sites are very limited in the vicinity of an amputation, the controllable degrees of freedom of an electric arm are few in number. It is possible, of course, to use other body sites and electromyographic electrodes for more sophisticated control signals. After all, the amputee would like to have an arm approaching the versatility of the normal human arm. In the IBM project, three pressure switches were installed in a pad worn in the shoe. The big toe, little toe, and heel could close these switches in various combinations to actuate various degrees of freedom. While the proper switch sequences were easily learned, control of the arm required excessive concentration by the wearer. Today's electric arms usually use a few pressure switches that can be activated inconspicuously.

Most amputees prefer the simplest prosthesis they can find and many will dispense with an artificial arm altogether rather than try to cope with a maze of wires and switches. Electrical artificial limbs can certainly be made simpler because modern technology has generated miniscule, logic circuits that can relieve the amputee of many control problems, particularly if EMG signal sources are used. A form of "supervisory control" in which various EMG signals from several body sites are blended electronically could give amputees almost natural control over their artificial limbs. This technological "fallout" from computer control work may be one of the important byproducts of space research.

One of the most sophisticated applications of EMG occurs in the "Boston Arm," designed by M. J. Glimcher of Harvard and R. W. Mann, an MIT professor. This electrically actuated prosthesis is controlled by signals from EMG electrodes on the biceps and triceps muscles on the stump. The actual hardware for the Boston Arm was developed at Liberty Mutual Insurance Company's Research Center, at Hopkinton, Massachusetts, under A. L. Cudworth.

## Electrical Master-Slaves

Mechanical master-slaves are undeniably extremely dexterous and versatile industrial manipulators. Their ability to operate through barriers and over large distances is limited by the lengths of their control cables. The bundle of control cables can be replaced by hardwire or radio links if electric servo motors are installed at both master and slave ends. Ray Goertz and his associates at Argonne National Laboratory accomplished this "electrification" of the master-slave in 1954. Without question, the ANL electrical master-slaves are superb examples of advanced teleoperator art. Only the cost and complexity of the electrical master-slave have retarded many commercial applications. In outer space and in some nuclear and undersea tasks, it is one of the best engineering solutions to the problem of projecting man's dexterity over distance and through recalcitrant barriers.

Argonne National Laboratory has built four different models of electrical master-slaves in the last decade and a half. Models E1 and E2 were developmental models. Four Model-E3 arms installed in the Chemical Engineering Senior Cave at Argonne have performed well for several years. The Mark E4A is presently a developmental model with such improvements over Model E3 as controllable force-multiplication ratios up to 5:1, lower cost, lower maintenance requirements, lower backlash and inertia, and better working geometry (Goertz, 1966).

The control circuits and servo arrangement of Mark E4A were described in Chapter 4. The motions and degrees of freedom of the Mark E4A are essentially the same as those of the Mod-8 mechanical master-slave. Servo drives and force-reflecting servos make the E4A completely bilateral. Most of the degrees of freedom are driven by tapes like those employed in mechanical master-slaves. The difference, of course, is that these E4A tapes are actuated by servo drive motors located in the rather substantial "body" of the slave arm. The entire slave "body" and arm are free to move in space as long as a hardwire or radio link is maintained with the master. Even terrestrially such mobility is an advantage. The E3 hot-cell installations at Argonne, for example, use bridge-crane type carriages to move the arms over large cell volumes, something impossible with mechanical master-slaves. Here is a case where the possible motions include seven bilateral degrees of freedom and five unilateral, switch-controlled degrees of freedom.

During the MSFC Independent Manned Manipulator study (discussed in Chapters 2 and 4), ANL investigated the possibilities of employing electric manipulators for the Maneuvering Work Platform (MWP) and Space Taxi concepts. Both the MWP and Space Taxi designs carried simple unilateral manipulators for docking and anchoring purposes. These arms would not be able to carry out dexterous operations in space. The Space Taxi concept also incorporated a pair of bilateral electric arms (Fig. 5.18). Each of the slave arms had seven master-slave degrees of freedom and eight indexing motions. Four of the indexing motions were intimately associated with master-slave degrees of freedom; they were used to expand the working volume accessible to the cramped master controls in the space capsule. When the envelope of the operator's control volume was reached, the slave arms were automatically repositioned. The other four indexing motions were switch-controlled and were employed to reposition grossly or "reshape" the slave-arm configurations. The Space Taxi manipulators employed the same servo and control tech-

| *Manipulator Specification* | *Suggested Value* |
|---|---|
| Configuration | Two 6-degree-of-freedom arms |
| Type | Bilateral (i.e., closed loop position control with force feedback) |
| Reach | 40-inch reach, spherical envelope |
| Response | Slightly less than man's response (about 4 cps bandwidth) |
| Resolution | 0.04 inch |
| Force | About 15 lb per arm minimum |
| End effector | Parallel-tong jaws |
| Video | Monocular (2 cameras) 1 fixed with pan and tilt 1 positionable by manipulator |
| Life | Approximately 10 days in orbit |
| Tethering and docking | Should allow easy repositioning of manipulator spacecraft |
| Indexing | Two shoulder joints |

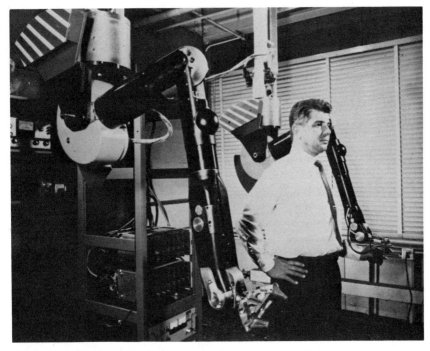

**Figure 7.10** The Brookhaven electric master-slave arm. The slave arm is on the right. Compare with the ANL E4A, Fig. 5.16. (Courtesy of C. R. Flatau.)

niques discussed in connection with the ANL E3 and E4A electric master-slaves.

The indexing techniques used in the Space Taxi concept are applied to all types of manipulators. One would expect that indexing would make master-slave operations difficult because it destroys spatial correspondence, but this factor becomes important only when the indexed angles exceed about 30°. In space and undersea applications, where manipulator control volumes are very restricted, indexing or some form of replica control must be adopted to gain reasonable working volume.

The only other space manipulators studied in detail were those designed for the General Electric repair and maintenance satellite. The GE work drew heavily on ANL electric master-slave experience as the manipulator specifications demonstrate on page 212.

The only other electric master-slave manipulator actually built is the unit designed by Flatau and his group at Brookhaven National Laboratory (Flatau, 1969). Built for mobile service in large accelerators, where residual radioactivity is high, the Brookhaven master-slave possesses only

## 214   The Actuator Subsystem

6 degrees of freedom (Fig. 7.10). Dc servo motors were used instead of the ac servos employed at ANL, resulting in size reduction. Rather than mounting all the servos in the shoulders and conveying motion by long tapes and cables (as was done in the ANL models), the Brookhaven manipulator has the actuators mounted on moving parts of the manipulator arm: two in a shoulder box and the remaining four in the lower arm. Consequently, the Brookhaven manipulator arm appears to be more bulky, but the shoulder box is more compact.

## ADVANCED ACTUATOR CONCEPTS

Electrical and hydraulic motors and pistons are convenient enough for most industrial and hostile-environment applications, but they are heavy, awkward, power-consuming, and often noisy. The deficiencies of conventional actuators have led to several studies of "artificial muscles."

Most of these investigators have tried to obtain linear force and motion through the surface distortion of flexible tube-like structures using gas pressure. The McKibben muscle, for example, consists of a straight braided sleeve and a gas-tight inner tube. When valves admit a gas or liquid, the cylinder bulges and the two ends are pulled together (U.S. Government, 1961). Other investigators have employed fiber glass tension fibers in an elastomer tube to achieve the same effect (Baldwin, 1963). Still another variation is the spherical-cell muscle studied by Reswick (Reswick, 1963). In this artificial muscle a rubber and cloth tube was constricted by bands at intervals along its length—looking something like a string of link sausages. When gas pressure is applied, each link or "cell" distorts and becomes more spherical. The overall effect is contraction, just as if a series of McKibben muscles were connected in series.

A slightly different tack was taken by B. F. Goodrich in a "rubber muscle" project. If a straight piece of rubber hose with specially wound reinforcing cord is pressurized with a liquid or gas it will bend to form an arc; if more pressure is applied, the curvature increases until the hose becomes a ring. Goodrich made a six-finger "hand" from this special hose that had some prehensile properties.

One wonders whether magnetic and electrostatic forces might not be employed to construct muscles more sophisticated than those made from deforming surfaces. The phenomena of magnetostriction and electrostriction do not provide enough contraction per unit length to serve as actuators. Electrostatic forces can be harnessed in principle to provide contraction, but the problems connected with the generation and safe handling of high voltages are very imposing, particularly in prosthetics and hostile environments. In fact, it is electromagnetic machinery rather than electrostatic machinery that dominates electric power technology. It is not surprising, then, to find electric muscles based on electromagnetic rather than electrostatic forces.

Rubber magnets frequently are used in refrigerator door latches, magnetic zippers, and the like. The core materials consist of rubber impregnated with magnetic particles. These are permanent magnets. By winding a coil solenoid-fashion about a dispersion of magnetic particles dispersed in soft rubber, the particles, which are tiny magnets, can be forced to attract one another and thus cause the rubber to contract. To provide a complete magnetic circuit, a magnetic muscle "cell" might be built in toroidal form. A current in the toroid windings creates a contractive force. Giannini Controls Corp. has constructed working models of magnetic muscles.

Going one step further, the impregnated rubber of the magnetic muscle may be replaced by a magnetic fluid, such as magnetic particles suspended in kerosene (Rosenzweig, 1966). Although still in the research-and-development stage, the magnetic muscle holds some promise for handicapped persons.

## TERMINAL DEVICES

If teleoperator hands were truly close approximations of human hands, hammers, saws, pliers, and other common tools could be used without modification. Teleoperator hands, however, will not be dexterous enough to handle these tools proficiently for many years. Tool-handling deficiencies now are partially remedied in three ways:

1. The rather crude, general-purpose teleoperator hands are replaced by specially designed tools that attach to the teleoperator wrists or, more often, by off-the-shelf tools modified so that they can be handled effectively by general-purpose hands.
2. The teleoperator is specially designed for easy tool interchange. Generally, this means that a rack must be supplied from which the teleoperator arm can pick up and replace tools in the proper orientation.
3. The task is designed with an eye to manipulator capabilities and limitations. Insistence upon captive nuts and bolts and special fixtures for holding dismantled parts are examples of such foresightedness.

The more specialized the task the more foresighted one can be, and the more specialized and effective one can make the hand and tool combinations. Action in emergencies, one of the teleoperator stocks in trade, cannot be thought out with as much precision, however, as the dismantling of the NERVA nuclear rocket engine. If teleoperators are to achieve their full potential, the hands and tools that do the work must be designed with great care.

A terminal device is whatever is at the end of a teleoperator wrist. It is the physical interface between the teleoperator and the task itself. It may be either a general-purpose "hand" or a special-purpose "tool."

The word "hand" applies to both the human-looking artificial hands that terminate many prostheses and the simple vise-like jaws on master-slave manipulators. The vise-type or parallel-jaw hand is by far the most common. As the jaws (also called "tongs" or "fingers") move toward or away from one another, they maintain their parallel relationship. Rods and other round objects twist easily between these plane surfaces. To prevent this, the jaws are sometimes notched or padded with a resilient material. In most hands, the jaws or fingers are remotely interchangeable.

The second major type of manipulator hand is the "hook" or "hook-and-anvil hand." Most unilateral manipulators use these interchangeably with the parallel-jaw hand. The preferred hook-type hand has a stationary anvil and movable notched "finger." In the unilateral electric manipulator, an electric motor pulls the finger toward the anvil.

Some jaws close like scissors, but flat jaws then are usually replaced by curved fingers similar to old-fashioned ice tongs. This type of hand is called a "grapple" or "claw" and is used most frequently in undersea work.

The "clamshell scoop" is also of marine origin, having been installed on Cousteau's *Diving Saucer*. The clamshell scoop opens and closes like a ring of flower petals. For this reason, it is also called a "petal," "blossom," or "orange-peel" hand. A clamshell scoop can grasp objects as well as gather samples of mud and sand.

**Terminal Tools**

Tools can be plugged into a manipulator wrist to replace the general-purpose hand. Obviously, these tools must be specially constructed to mesh with the fittings, gears, and drive shafts of a particular teleoperator wrist. Specialization makes them expensive but more effective in narrow lines of work.

Hand-held tools comprise another class. How can a choice be made between the two types of tools for a given application? Some considerations are:

1. Is the hand strong enough to handle the contemplated tool? If not, a specially designed, wrist-attached tool may be lighter than a hand-held

general purpose tool. The wrist joint can doubtless handle more weight than the fingers of a hand. The same kinds of considerations apply to power, torque, and grip forces.

2. Can the power be conveniently switched on and off? When the power comes from the manipulator wrist, the switch is built into the manipulator controls.

3. In an emergency, could the wrist-attached hand be readily exchanged for a general purpose hand?

4. Is there a chance that the wrist-attached tool might get stuck or somehow wedged in the work so that the arm could not be retracted?

5. Which approach will get the job done better in terms of time, cost, and other mission figures of merit?

In the nuclear industry, where these questions first arose, the hand-held tool is favored. Part of this preference is because the great bulk of hot-cell manipulators are mechanical master-slaves that have no motor drives at the slave wrist. Indeed, the question of the tools required for a mission may help determine the kind of manipulator finally selected. If the mission involves a great deal of bolting and unbolting, an electric or hydraulic arm with a special wrench replacing the hand may be more effective than a power wrench held by a master-slave. The force-reflecting master-slaves are usually superior in the matter of tool manipulation and control. For example, sawing without a sense of feel might lead to saw binding and breakage.

If the decision is in favor of a wrist-attached tool, a rather good selection of tools is commercially available. Others can be readily built from proven designs, particularly for unilateral manipulators.

## Hand-Held Tools

Hand-held tools may be specially built for manipulator use, or commercially available tools modified slightly to make them easier to handle with the single degree of freedom available in the teleoperator hand.

Some of the simpler tools such as pliers may be permanently fitted with manipulator fingers. Such an assembly is termed an "integral hand-tool combination." It has the disadvantage of requiring finger changes each time the tool is used.

More common and more versatile is the "adapter block." An adapter block is simply a chunk of metal with finger slots milled on two sides for grasping with the vise type of manipulator hand. The adapter block can be permanently attached to many tools, such as saws, grinders, and even radiation meters.

A good tool rack promotes effective tool use, especially when one is working with a single arm in space or under the ocean. Tools should be stored in a position in which the manipulator hand can firmly grasp them, extract them from the rack, and replace them. A dropped tool in a hostile area may be a lost tool.

## Task Design

If a machine, say a nuclear rocket engine, is being disassembled, each piece must be designed so that the manipulator can (1) reach it, (2) unfasten it easily, (3) grab it firmly, (4) extract it and lift it clear, and (5) set it down in a position and attitude that permit easy recovery (Morand, 1961). Naturally, the same points apply to assembly, but in reverse order.

A number of mechanical devices are available to help the manipulator operator in assembly-disassembly sequences. In the first category are such things as captive nuts and bolts that cannot fall to the floor and be lost. They also promote reassembly. Many electrical and hydraulic "quick disconnects" are amenable to remote handling. In the second category are the so-called "fixtures." These are special stands that hold the manipulator-deposited parts in the proper positions for manipulator recovery. Finally, there are the guide pins and grooves that materially aid the operator in correctly positioning components during reassembly operations. All of these strategens require the foreknowledge and cooperation of the machine's designer long before remote operations begin.

# VIII
# CONCLUSIONS AND FORECAST

Thousands of teleoperators have been built and used successfully in handling radioactive materials, helping the handicapped, and working on the ocean bottom. But these teleoperators are poor and incomplete extensions of man, with only a small fraction of man's dexterity and man's many degrees of freedom. Such is the current state of the art, but our survey has noted many scattered harbingers of growth. This growth will meet demands that man cannot fulfill without machine aid and it is being encouraged by many new technical developments.

Such developments are seldom breakthroughs when taken separately. Together, however, recent advances are giving us the ability to build a new generation of teleoperators. Their subsystems are benefitting directly from aerospace and related technology as indicated below:

It is tempting to predict which industry—and men, as a consequence—will benefit most from improved teleoperators. But radically new developments have a habit of becoming valuable where least expected.

Some "fallout" is highly probable in the prosthetics field as engineers begin to apply new materials, better power supplies and control techniques. Public services may increasingly need teleoperators to handle the more dangerous byproducts of our civilization. But these are down-to-earth and rather conservative thoughts.

Plans for harnessing teleoperators need not and must not be limited by today's crude mechanical arms with their few degrees of freedom or by today's primitive walking machines and exoskeletons. The man-controlled teleoperator enables man to conquer distance, high temperatures, high pressures, noxious atmospheres, and other recalcitrant environments on the periphery of his narrow domains.

Scientific gadgetry may some day project a human being to wherever he wants to be and faithfully duplicate that spot's environment as

| Teleoperator Subsystem | Developments Benefitting Teleoperators |
|---|---|
| Actuator subsystem | Miniature motors, magnetic muscles, stepping motors |
| Sensor subsystem | Miniature TV cameras, tactual sensors, sonar imagers, infrared devices |
| Control subsystem | Digital control techniques, computer-generated visual displays, computer control systems (supervisory control), EMG |
| Communication subsystem | PCM refinements, lasers, miniature equipment |
| Computer subsystem | Fast, lightweight computers and memories |
| Power subsystem | Miniature batteries, lightweight solar and nuclear power plants |
| Environment-control subsystem | Space life-support systems |
| Structure subsystem | Strong lightweight materials |

well as the operator's actions. One can even conceive of a great surgeon operating on a patient a thousand miles away via a teleoperator with great dexterity and acute tactual feedback. The augmentation and extension of man by teleoperator will also help tap new lodes of raw materials and food supplies, such as those now locked in the deep oceans. A teleoperator can place the surface of Mars or the ocean floor at the scientist's fingertips. Conceivably, man-machine symbiosis can make a man a superman, either on the spot he occupies or on the other side of the universe.

# BIBLIOGRAPHY

Adams, J. A., and C. Webber, Monte Carlo Model of Tracking Behavior, *Human Factors,* Vol. 5, 1963, pp. 81-102.

Adams, J. J., and H. P. Bergeron, *Measurements of the Human Transfer Function with Various Model Forms,* NASA TN D-2394, 1964.

Adams, J. J., *A Simplified Method for Measuring Human Transfer Functions,* NASA TN D-1782, 1963.

Adams, J. J., H. P., Bergeron, and G. J. Hurt, *Human Transfer Functions in Multi-Axis and Multi-Loop Control Systems,* NASA TN D-3305, 1966.

Adams, J. L., *An Investigation of the Effects of the Time Lag Due to Long Transmission Distances Upon Remote Control,* NASA TN D-1211, 1961.

Adams, J. L., *An Investigation of the Effects of the Time Lag Due to Long Transmission Distances Upon Remote Control,* Phase II, NASA TN D-1351, 1962.

Ahrendt, W. F. and C. J. Savant, Jr., *Servomechanism Practice.* McGraw-Hill Book Co., New York, 1960.

Akulinicher, I., *Bioelectric Control,* AD-672910, 1967.

Alles, D. S., Kinesthetic Feedback System for Amputees via the Tactile Sense, Sc. D. thesis, M.I.T., 1968.

Alter, R., *Bioelectric Control of Prosthesis,* AD-646218, 1966.

American Institute of Chemical Engineers, *Proceedings of the Sixth Hot Laboratory and Equipment Conference,* AEC TID-7556, 1958.

American Nuclear Society, *Proceedings of the Eighth Conference on Hot Laboratories and Equipment,* AEC TID-7599, 1960.

American Nuclear Society, *Proceedings of the Ninth Conference on Hot Laboratories and Equipment,* Chicago, 1961.

American Nuclear Society, *Proceedings of the Tenth Conference on Hot Laboratories and Equipment,* Chicago, 1962.

American Nuclear Society, *Proceedings of the Eleventh Conference on Hot Laboratories and Equipment,* Hinsdale, 1963.

American Nuclear Society, *Proceedings of the Twelfth Conference on Remote Systems Technology,* Hinsdale, 1964.

American Nuclear Society, *Proceedings of the 13th Conference on Remote Systems Technology,* Hinsdale, 1965.

American Nuclear Society, *Proceedings of the 14th Conference on Remote Systems Technology,* Hinsdale, 1966.

American Society of Mechanical Engineers, *Proceedings of the Seventh Conference on Hot Laboratories and Equipment,* New York, 1959.

# Bibliography

Anderson, M. H., Harness and Control Systems, *Manual of Upper Extremity Prosthetics*, W. R. Santschi, ed., University of California, Los Angeles, 1958, pp. 155-187.

Anderson, V. C., *MPL Experimental RUM*, Marine Physical Laboratory, Scripps Institution of Oceanography, SIO Ref. 60-26, 1960.

Anderson, V. C., Vehicles and Stations for the Installation and Maintenance of Sea Floor Equipment, *IEEE Spectrum*, Vol. 1, November 1964.

Anderson, V. C., *Underwater Manipulators in the Benthic Laboratory Program of the Marine Physical Laboratory*, American Nuclear Society paper, 1964.

Anderson, V. C., and H. A. O'Neal, *Manipulators and Special Devices*, AD-609490, 1964.

Anderson, V. C., Maintenance of Sea-Floor Electronics, *Transactions IEEE*, Vol. AES-21, September 1968, p. 650.

Anderson, V. C., and R. C. Horn, *Tensor Arm Manipulator Design*, ASME Paper 67-DE-57, 1967.

Anon., Hot Laboratory and Special Handling Equipment, *Atomics*, Vol. 17, May-June 1964, pp. 19–30.

Anon., "Adelbert"—Science's "Right Arm"—Can Even Write Its Name, *Popular Mechanics*, Vol. 96, October 1951, p. 161.

Anon., Undersea Remote Technology, *Proceedings of the Twelfth Conference on Remote Systems Technology*, American Nuclear Society, Hinsdale, 1964, pp. 279-282.

Anon., General Electric Power-Operated Manipulator, *Engineering*, Vol. 185, March 1958, p. 397.

Anon., Almost-Human Engineering, *Machine Design*, Vol. 31, April 30, 1959, pp. 22-26.

Anon., Human Engineering in Remote Handling, U.S. Air Force, MRL-TDR-62-58, 1962.

Anon., Dummies with Muscles Will Pass Judgment on Spacesuit, *Machine Design*, Vol. 35, December 5, 1963, p. 6.

Anon., Develop Missile Loader for Navy, *Product Engineering*, Vol. 30, June 1, 1959, p. 22.

Anon., Mobile Protection for Explosive Handlers, *Missiles and Rockets*, Vol. 11, July 16, 1962, p. 36.

Anon., *Proceedings of a Symposium on Powered Prostheses*, The Working Party on Powered Limbs and Related Appliances of the Minister of Health's Standing Advisory Committee on Artificial Limbs, October 29, 1965.

Anon., Extra-Maneuverable Manipulators, *Product Engineering*, Vol. 26, May 1955, pp. 140-143.

Anon., Mechanical Slave Performs at Master's Bidding, *Electrical Engineering*, Vol. 75, 1956, p. 671.

Anon., Hydraulics Tries Its Hand, *Machine Design*, February 29, 1968, pp. 24-26.

Anon., *External Control of Human Extremities*, Yugoslav Committee for Electronics and Automation, Belgrade, 1967.

Anon., *Digest of the 7th International Conference on Medical and Biological Engineering*, Almqvist and Wiksell, Stockholm, 1967.

Anon., The Humanoids Are Coming to Do the Dirty Work, *Product Engineering*, Vol. 38, August 1967, pp. 30-32.

Anon., The Servomanipulator Mascot III, *Comitato. Nationale per l'Energia Nucleare, Notiziario*, Vol. 14, 1968, pp. 58-59.

Anon., no title, *Life*, May 3, 1948.

Argonne National Laboratory, *A Manual of Remote Viewing*, ANL-4903, 1952.

Argonne National Laboratory, *Manipulator Systems for Space Application*, Technical Report, Argonne, 1967.

Arnold, J. E., and P. W. Braisted, *Design and Evaluation of a Predictor for Remote Control Systems Operating with Signal Transmission Delays*, NASA TN D-2229, 1963.

Arp, H., *Some Design Aspects of an Experimental Fluidic Control System*, PB-180954, 1968.

Arzebaecher, R. C., *Servomechanisms with Force Feedback*, ANL-6157, 1960.

Athans, M., and P. L. Falb, *Optimal Control*, McGraw-Hill Book Co., New York, 1966.

Atomic Energy Commission, *Proceedings of the 1964 Seminars on Remotely Operated Special Equipment*, AEC CONF-640508 and AEC CONF-641120, 1964.

Auerbach Corp., *Visual Information Display Systems*, NASA SP-5049, 1968.

Ball Brothers, Inc., *Experiments for Satellite and Material Recovery from Orbit (ESMRO) Study Program*, Boulder, November 8, 1966.

Baker, D. F., compiler, *Survey of Remote Handling in Space*, U. S. Air Force AMRL-TDR-62-100, 1962.

Baldwin, H. A., et al., *Study and Development of Muscle Substitutes*, U. S. Air Force RTD-TDR-63-4181, AD-431825, 1963.

Ballinger, H. A., Machines with Arms. *Science Journal*, Vol. 4, October 1968, pp. 59-65.

Barabaschi, S., et al., An Electronically Controlled Servo-Manipulator, *Proceedings of the Ninth Hot Laboratory and Equipment Conference*, American Nuclear Society, Chicago, 1961, pp. 143-153.

Barber, D. J., *Mantran: A Symbolic Language for Supervisory Control of an Intelligent Remote Manipulator*, NASA CR-88271, 1967.

Bates, J. K., A Classification of Information Display, *Information Display*, Vol. 3, March-April 1966, pp. 47-51.

Bayerskiy, R. M., and V. V. Parin, *Introduction to Medical Cybernetics*, NASA TT-F-459, 1966.

Beach, F. J., A Flexible and Versatile Display for Command and Control, *Information Display*, Vol. 4, May-June 1967, pp. 62-66.

Becher, A. F., *A Mobile Manipulator for Use in Controlling Radiation Emergencies*, Union Carbide Rep. K-C-768, 1965.

Beckers, R. M., et al., *Engineering Design Practices at ORNL for Facilities Containing Radioactive Materials*, ORNL-TM-1459, 1966.

Beckett, J. T., *A Computer-Aided Control Technique for a Remote Manipulator*, NASA CR-88483, 1967.

Bekey, G. A., The Human Operator as a Sampled Data System, *Transactions IRE*, Vol. HFE-3, September 1962, pp. 43-51.

Bekey, G. A., H. F. Meissinger, and R. E. Rose, Mathematical Models of Human Operators in Simple Two-Axis Manual Control Systems, *Transactions IEEE*, Vol. HFE-6, September 1965, pp. 42-52.

Bekey, G. A., *An Investigation of Sampled Data Models of the Human Operator in Control System*, U. S. Air Force ASD-TDR-62-36, 1962.

Bekey, G. A., *Research on New Techniques for the Analysis of Manual Control Systems*, NASA CR-89592, 1967.

Bekker, M. G., *Introduction to Terrain Vehicle Systems*, Silver Burdett Co., Morristown, 1968.

Bekker, M. G., *Theory of Land Locomotion*, University of Michigan Press, Ann Arbor, 1956.

Bekker, M. G., *Off-the-Road Locomotion*, University of Michigan Press, Ann Arbor, 1960.

Bennet, W. F., *Some Man-Machine Problems in Remote Handling Equipment*, U. S. Air Force AFSWC TN 59-24, 1959.

Bennett, E., ed., *Human Factors in Technology*, McGraw-Hill Book Co., New York, 1963, pp. 425-443.

Beno, J. H., *Mobot the Robot*, SAE Paper 570A, 1962.

Beno, J. H., and D. A. Campbell, New Devices for Deep Sea Operations, *Undersea Technology*, Vol. 4. 1963, p. 25.

Berbert, A. G., and C. R. Kelley, Piloting Nuclear Submarines with Controls that Look Into the Future, *Electronics*, Vol. 35, June 8, 1962, pp. 35-39.

Bergeron, H. P., J. K. Kincaid, and J. J. Adams, *Measured Human Transfer Functions in Simulated Single-Degree-of-Freedom Nonlinear Control Systems*, NASA TN D-2569, 1965.

Berlanger, P. R., *Time-Varying Characteristics of the Human Operator in an Open Loop*, AD-275774, 1962.

Bernotat, R., and H. Widlok, Principles and Applications of Prediction Display, *Journal of the Institute of Navigation*, Vol. 19, July 1966, pp. 361-370.

Bernotat, R., *Anthropotechnique as a Scientific Discipline*, NASA TT-F-11390, 1966.

Bert, J. D., Jr., *Visual and Audio Feedback for Mobile Remote Operations*, AD-447107, 1967.

Bertone, C. M., *A Bibliography of Russian Scientific and Technological Literature In Manual Control and Associated Areas*, NASA CR-199, 1965.

Bever, J. G., et al., *Development of Visual, Contact and Decision Subsystems for a Mars Rover*, NASA CR-87503, 1967.

Bevilacqua, F., *Force Reflecting Servomechanism*, AEC TID-10074, 1951.

Birmingham, H. P., The Optimization of Man-Machine Control Systems, *IRE WESCON Convention Record*, Part 4, August 1958, pp. 272-276.

Birmingham, H. P., and F. V. Taylor., *A Design Philosophy for Man-Machine Control Systems*, Proceedings IRE, Vol. 42, December 1954, pp. 1748-1758.

Blackmer, R. H., A. Interian, and R. G. Clodfelter, *The Role of Space Manipulator Systems for Extravehicular Tasks*, General Electric Co., Schenectady, 1968.

Blackmer, R. H., and R. G. Clodfelter, The Application of Remote Manipulators in Space, *Transactions ANS*, Vol. 11, November 1968, pp. 617-618.

Blackmer, R. H., ed., *Remote Manipulators and Mass Transfer Study, Final Report*, General Electric Co., Schenectady, 1968.

Bliss, J. C., and H. D. Crane, *Experiments in Tactual Perception*, NASA CR-322, 1965.

Bliss, J. C., J. W. Hill, and B. M. Wilber, *Characteristics of the Tactile Information Channel*, NASA CR-1389, 1969.

Bliss, J. C., and J. W. Hill, *Tactile Perception Studies Related to Control Tasks*, Progress Reports on NASA Contract, 1967-1968.

Bliss, J. C., *Human Operator Describing Functions with Visual and Tactile Displays*, In NASA SP-144, March 1967, pp. 67-79.

Bliss, J. C., et al., Information Available in Brief Tactile Presentations, *Perception and Psychophysics*, Vol. 1, 1966, pp. 273-283.

Bliss, J. C., and H. D. Crane, Communicating Through the Human Integument, *The Control of External Power in Upper Extremity Rehabilitation*, NASA/NRC 1352, National Academy of Sciences, Washington, 1966, pp. 255-261.

Boehm, B. W., Keeping the Upper Hand in the Man-Computer Partnership, *Astronautics and Aeronautics*, Vol. 5, April 1967, pp. 22-28.

Boone, A. R., Plug-In Workman Built in 90 Days, *Popular Science*, Vol. 163, December 1953, pp. 100-103.

Booth, T. L., et al., *Experimental Investigations of Man-Machine Processing of Information*, AD-653278, 1966.

Bost, W. E., *Hot Laboratories, An Annotated Bibliography*, AEC TID-3545, revision 1, 1965.

Bottomley, A. H., and T. R. Cowell, An Artificial Hand Controlled by the Nerves, *New Scientist*, No. 382, March 12, 1964, pp. 668-671.

Bottomley, A. H., A. B. Kinnier-Wilson, and A. Nightingale, Muscle Substitutes and Myoelectric Control, *Journal British IRE*, Vol. 26, December 1963, pp. 439-448.

Bottomley, A. H., Progress with the British Myoelectric Hand, *External Control of Human Extremities*, Yugoslav Committee for Electronics and Automation, Belgrade, 1967, pp. 114-124.

Bottomley, A. H., Signal Processing in a Practical Electromyographically Controlled Prosthesis, *The Control of External Power in Upper Extremity Rehabilitation*, NAS/NRC 1352, National Academy of Sciences, Washington, 1966, pp. 271-273.

Bradley, W. E., *Servomechanism-Propelled Vehicles*, AD-653835, 1967.

Bradley, W. E., *Telefactor Control of Space Operations*, AIAA Paper 66-918, 1966.

Braman, H. R., ed., *Human Factors of Remote Handling in Advanced Systems*, U. S. Air Force ASD TR 61-430, 1961.

Bricker, L., Working in Space: Are we Ready? *Space/Aeronautics*, Vol. 46, October 1966, pp. 68-76.

Briscoe, G. J., et al., *Preliminary Technical Review—Nuclear Accident Recovery Equipment*, General Electric GEMP-394, 1965.

Brown, J. A., and W. A. Koelsch, Jr., *A Compact Mobile Manipulator*, AEC TID-7599, 1960, pp. 224-229.

Brown, J. E., *Preliminary Report on Advanced Manipulator Studies*, General Electric DC-60-7-137, 1960.

Brown, J. E., *Preliminary Report of Advanced Viewing Studies for Remote Handling Operations*, General Electric DC-60-8-32, 1960.

Burnett, J. R., Force-Reflecting Servos Add Feel to Remote Controls, *Control Engineering*, Vol. 4, July 1957, pp. 82-87.

Burnett, J. A., R. C. Goertz, and W. M. Thompson, Mechanical Arms Incorporating a Sense of Feel for Conducting Experiments with Radioactive Materials, *Proceedings of the International Conference on Peaceful Uses of Atomic Energy*, Vol. 14, Geneva, 1955, p. 116.

Burrows, A. A., Control Feel and the Dependent Variable, *Human Factors*, Vol. 7, October 1965, pp. 413-421.

Burton, J. H., *The SNAP Environmental Test Facility*, ARS Paper 1650-61, 1961.

Burton, J. H., *SETF Remote Viewing Techniques*, AEC TID-7599, 1960, pp. 263-276.

Callery Chemical Co., *Design of Safety Equipment for Handling High-Energy Research Materials of Unknown Sensitivity*, AD-263378, 1961.

Campbell, D. A., Multiplex Circuits for Circuits of Robots, *Electronics*, Vol. 33 January 22, 1960, pp. 46-48.

Carlson, W. D., The GE MAN II Master-Slave Manipulator, *Proceedings of the Fourth Hot Laboratory and Equipment Conference*, AEC, 1955, pp. 11-25.

Cautin, W. J., and F. Rapp, *Description of Easy English*, AD-660569, 1967.

Chalmers, W. J., Force-Reflecting Servomechanism Research—Servomechanism Synthesis Through Network Parameters, Ph. D. thesis, Purdue University, Lafayette, 1955.

Charron, A. G., *Remote Man-Machine Control System Evaluation, Final Report*, NASA CR-76889, 1964.

Chidambara, M. R., et al., *Continuation of Theoretical and Experimental Research on Digital Adaptive Control Systems*, NASA CR-810, 1967.

Chubb, G. P., An Evaluation of Proposed Applications of Remote Handling in Space, *Proceedings of the 1964 Seminars on Remotely Operated Special Equipment*, Vol. 1, AEC CONF-640508, 1964, pp. 233-239.

Clark, D. C., *Exploratory Investigation of the Man-Amplifier Concept*, U.S. Air Force AMRL-TDR-62-89, AD-390070, 1962.

Clark, D. C., N. J. Deleys, and C. W. Matheis, *Exploratory Investigation of the Man-Amplifier Concept*, U.S.A.F. AMRL-TDR-62-89, 1962.

Clark, D. C., et al., *Exploratory Investigation of the Man Amplifier Concept*, AD-290070, 1962.

Clark, J. W., *Unmanned Ground Support Equipment*, U.S.A.F. ASD TR 61-430, 1961, pp. 43-63.

Clark, J. W., Mobotry: New Art of Remote Handling, *Transactions IRE*, Vol. VC-10, August 1961, pp. 12-24.

Clark, J. W., The Mobot Mark II Remote Handling System, *Proceedings of the Ninth Hot Laboratory and Equipment Conference*, American Nuclear Society, Chicago, 1961, pp. 111-120.

Clark, J. W., Telechirics—for Operations in Hostile Environments, *Battelle Technical Review*, Vol. 12, October 1963, pp. 3-8.

Clark, J. W., Analysis of Hostile-Environment Methodologies, *Proceedings of the 1964*

*Seminars on Remotely Operated Special Equipment*, Vol. 1, AEC CONF-640508, 1964, pp. 4-23.

Clark, J. W., A Taxonomy for Remotely Operated Systems, *Proceedings of the Twelfth Conference on Remote Systems Technology*, American Nuclear Society, Hinsdale, 1964, pp. 105-116.

Colgan, J. E., Jr., Architect-Engineering Considerations in the Design of Remote Handling Tools as to Function and Flexibility, *Proceedings of the 1964 Seminars on Remotely Operated Special Equipment*, Vol. 1, AEC CONF-640508, 1964, pp. 340-362.

Comeau, C. P., and J. S. Bryan, Headsight Television System Provides Remote Surveillance, *Electronics*, November 10, 1961, pp. 86-90.

Conklin, J. E., Effect of Control Lags on Performance of a Tracking Task, *Journal of Experimental Psychology*, Vol. 53, 1957, pp. 261-268.

Control Data Corporation, *Foveal Hat—A Head-Aimed Television System with a Dual Field of View*, Report TM-74-98, Rosemont, Pa., 1967.

Corliss, W. R., *Scientific Satellites*, NASA SP-133, 1967.

Corliss, W. R., and E. G. Johnsen, *Teleoperator Controls*, NASA SP-5070, 1968.

Craik, K. J. W., Theory of the Human Operator in Control Systems, *British Journal of Psychology*, Vol. 38, December 1947, pp. 56-61, and Vol. 38, March 1948, pp. 142-148.

Crawford, B. M., and W. N. Kama, *Remote Handling of Mass*, U.S.A.F. ASD TR 61-627, AD-273491, 1961.

Crawford, B. M., Joy Stick Versus Multiple Levers for Remote Manipulator Control, *Human Factors*, Vol. 6, February 1964, pp. 39-48.

Crawford, B. M., and D. F. Baker, *Human Factors in Remote Handling: Survey and Bibliography*, U.S.A.F. WADD TR-60-476, AD-242524, 1960.

Crawford, B. M., *Measures of Remote Manipulator Feedback*, U.S.A.F. WADD TR 60-591, 1961.

Cuffia, R. J., et al., *Accident Recovery Equipment Study AEC-DRD Reactors*, AEC IDO-10043, Rev. 1, Vols. 1 and 2, 1965.

Curtis, W. K., The Development of the Master-Slave Manipulator, *Nuclear Energy*, Vol. 4, November 1963, p. 1416.

Davis, J. M., *Operator Selection, Training, and Efficiency in the Field of Remote Handling*, U.S.A.F. ASD TR 61-430, 1961, pp. 11-17.

Desroche, M., and G. Cherel, Gas-Tight Cell and Magnetic Remote Controlled Manipulator, *Proceedings of the Ninth Hot Laboratory and Equipment Conference*, American Nuclear Society, Chicago, 1961, pp. 87-90.

Devaney, A. J., C. Grauling, and S. Baron, *Holographic Display Systems*, In NASA SP-144, March 1967, pp. 55-65.

Diamantides, N. C., Man as a Link in a Control Loop, *Electrotechnology*, Vol. 69, January 1962, pp. 40-46.

Drexler, R. L., and R. B. O'Brien, *Preliminary Technical Review, A.E.C. Vehicular Manipulating Systems*, General Electric GEMP-393, 1965.

Drexler, R. L., *Functional Requirements, Small Modular Recovery Vehicle System*, General Electric GEMP-417, 1966.

Dreyfus, H., *The Measure of Man: Human Factors in Design*, Whitney Library of Design, 1968.

Easterby, R. S., Perceptual Organization in Static Displays for Man/Machine Systems, *Ergonomics*, Vol. 10, March 1967, pp. 195-205.

Elkind, J. I., A Survey of the Development of Models for the Human Controller, *Guidance and Control-II*, R. C. Langford and C. J. Mundo, eds., Academic Press, New York, 1964, pp. 613-643.

Elkind J. I., and C. D. Forgie, Characteristics of the Human Operator in Simple Manual Control Systems, *Transactions IRE*, Vol. AC-4, May 1959, pp. 44-45.

Elkind, J. I., and D. C. Miller, *Process of Adaption by the Human Controller*, In NASA SP-128, March 1966, pp. 47-63.

Elkind, J. I., and L. T. Sprague, Transmission of Information in Simple Manual Control Systems, *Transactions IRE*, Vol. HFE-2, March 1961, pp. 58-60.

Engelberger, J. F., Enter, The Industrial Robot, *Industrial Research*, November 1968, pp. 56-61.

Ernst, H. A., MH1—A Computer Operated Mechanical Hand, Sc. D. thesis, M.I.T. 1961.

Falb, P. L., and G. Kovatch, *Dynamical System Modeling of Human Operators—A Preliminary Report*, In NASA SP-128, 1966, pp. 345-358.

Fargel, L. C. and E. A. Ulbrich, Aerospace Simulators Prove Predictor Displays Extend Manual Operation, *Control Engineering*, Vol. 10, 1963, pp. 57-60.

Farr, D. E., and D. Meister, *The Methodology of Control Panel Design*, AD-646442, 1966.

Feigenbaum, E. A., and J. Feldman, eds., *Computers and Thought*, McGraw-Hill Book Co., New York, 1967.

Ferguson, K. R., Design and Construction of Shielding Windows, *Nucleonics*, Vol. 10, November 1952, pp. 46-51.

Ferguson, K. R., W. B. Doe, and R. C. Goertz, Remote Handling of Radioactive Materials, *Reactor Handbook*, Vol. IV, Interscience Publishers, New York, 1964, pp. 463-538.

Ferrell, W. R., *Remote Manipulation with Transmission Delay*, NASA TN D-2665, 1965.

Ferrell, W. R., and T. B. Sheridan, *Measurement and Display of Control Information*, NASA CR-83980, 1966. One of a series of M.I.T. Progress Reports.

Ferrell, W. R., and T. B. Sheridan, Supervisory Control of Remote Manipulation, *IEEE Spectrum*, Vol. 4, October 1967, pp. 81-88.

Ferrell, W. R., Delayed Force Feedback, *Human Factors*, Vol. 8, October 1966, pp. 449-455.

Ferrier, M., ed., *Proceedings of the Twelfth Conference on Remote Systems Technology*, American Nuclear Society, Hinsdale, 1964.

Field, R. E., and J. F. Gifford, *Development of the Hanford Slave Manipulator for Use in the Multicurie Cells at Hanford*, AEC HW-26175, 1952.

Figenshau, J. K., *Manipulators for Nuclear and Other Hazardous Environments*, Paper 727B National Farm, Construction and Industrial Machinery Meeting, 1962, AEC CONF-191-6, 1962.

Fishlock, D., Four Finger Exercise, *New Scientist*, May 4, 1967.

Flatau, C. R., Development of Servo Manipulators for High Energy Accelerator Requirements, *Proceedings of the 13th Conference on Remote Systems Technology*, American Nuclear Society, Hinsdale 1965, pp. 29-35, AEC BNL-9388, 1965.

Flatau, C. R., *Servo Telemanipulators and Their Present and Future Applications*, BNL-13867, 1969.

Flatau, C. R., Compact Servo Master-Slave Manipulator with Optimized Communications Links, *Proceedings of the 17th Conference on Remote Systems Technology*, 1969.

Flatau, C. R., General-Purpose Servo-Manipulator for Remote Maintenance of Accelerators, *Transactions IEEE*, Vol. NS-16, June 1969, pp. 594-598.

Fletcher, M. J., and F. Leonard, The Principles of Artificial Hand Design, *Artificial Limbs*, May 1955, pp. 78-94.

Fletcher, M. J., *Some Considerations in the Design of Hand Substitutes*, ASME Paper 59-A-262, 1959.

Fogel, L. J., and R. A. Moore, *Modeling the Human Operator with Finite-State Machines*, NASA CR-1112, 1968.

Fogel, L. J., *Biotechnology: Concepts and Applications*, Prentice-Hall, Inc., Englewood Cliffs, 1963.

Fogel, L. J., *On the Design of Conscious Automata*, AD-644204, 1966.

Ford, A., and C. T. White, *The Effectiveness of the Eye as a Servo Control System*, U.S.N. Electronics Laboratory, Report 934, 1959.

Freedy, A., L. F. Lucaccini, and J. Lyman, *Information and Control Analysis of Externally Powered Artificial Arm Systems*, Biotech. Lab. Tech. Report No. 42, University of California, Los Angeles, 1967.

Freedy, A., L. F. Lucaccini, and J. Lyman, *An Adaptive Approach to Optimum Switching Control of Artificial Arms*, Report 67-47, University of California, Los Angeles, 1967.

Freeman, R. A., ed., *TAN Hot Shop and Satellite Facilities*, AEC IDO-17032, 1964.

Froelich, H. E., *Integrated Controls for Undersea Vehicle-Manipulator Systems*, Paper ASME Underwater Technology Conference, New London, 1965.

Frost G. G., et al., Man-Machine Dynamics, unpublished manuscript, 1968.

Frost, G. G., and W. K. McCoy, Jr., A "Predictor" Display for On-Board Rendezvous Optimization, *Proceedings National Electronics Conference*, Vol. 21, 1965, pp. 683-688.

Furman, B., *Progress in Prosthetics*, Government Printing Office, Washington, 1962.

Galbiati, L., et al., A Compact and Flexible Servosystem for Master-Slave Electric Manipulators, *Proceedings of the Twelfth Conference on Remote Systems Technology*, American Nuclear Society, Hinsdale, 1964, pp. 73-87.

Garner, K. C., Evaluation of Human Operator Coupled Dynamic Systems, *Ergonomics*, Vol. 10, March 1967, pp. 125-138.

General Electric Co., *Hardiman I Prototype Project, Special Interim Study*, Report S-68-1060, Schenectady, 1968.

General Electric Co., *Exoskeleton Prototype Project, Final Report on Phase I*, AD-807467L, Schenectady, 1968.

General Electric Co., *Special Technical Report on Joints in Series*, S-68-1081, Schenectady, 1968.

General Electric Co., *Application of Remote Manipulation to Satellite Maintenance*, NASA CR-73388 and CR-73389, 1969.

General Electric Co., *Remote Manipulators and Mass Transfer Study*, AFAPL-TR-68-75, 1968.

General Electric Co., *Machine Augmentation of Human Strength and Endurance*, Report S-69-1116, 1969.

Goertz, R. C., *Master-Slave Manipulator*, AEC ANL-4311, 1949.

Goertz, R. C., *Manipulator Philosophy and Development*, AEC TID-10074, 1951.

Goertz, R. C., and F. Bevilacqua, A Force Reflecting Positional Servomechanism, *Nucleonics*, Vol. 10, November 1952, pp. 43-45.

Goertz, R. C., Fundamentals of General-Purpose Remote Manipulators, *Nucleonics*, Vol. 10, November 1952, pp. 36-42.

Goertz, R. C., J. R. Burnett, and F. Bevilacqua, *Servos for Remote Manipulation*, AEC ANL-5022, 1953.

Goertz, R. C., Mechanical Master-Slave Manipulator, *Nucleonics*, Vol. 12, November 1954, pp. 45-46.

Goertz, R. C., and F. Bevilacqua, A Force Reflecting Positional Servo-Mechanism, *Nucleonics*, Vol. 10, November 1952, pp. 43-45.

Goertz, R. C., and W. M. Thompson, Electronically Controlled Manipulator, *Nucleonics*, Vol. 12, November 1954, pp. 46-47.

Goertz, R. C., and W. M. Thompson, *Proceedings of the Fourth Hot Laboratory and Equipment Conference*, 1955.

Goertz, R. C., et al., *The ANL Model 3 Master-Slave Electric Manipulator—Its Design and Use in a Cave*, AEC TID-13237, 1961.

Goertz, R. C., *Human Factors in Design of Remote-Handling Equipment*, U.S.A.F. ASD TR 61-430, 1961, pp. 169-172.

Goertz, R. C., Manipulators Used for Handling Radioactive Materials, *Human Factors in Technology*, E. Bennett, ed., McGraw-Hill Book Co., New York, 1963, pp. 425-443.

Goertz, R. C., Some Work on Manipulator Systems at ANL; Past, Present, and a Look at the Future, *Proceedings of the 1964 Seminars on Remotely Operated Special Equipment*, Vol. 1, AEC CONF-640508, 1964, pp. 27-69.

Goertz, R. C., Manipulator Systems Development at ANL, *Proceedings of the Twelfth Conference on Remote Systems Technology*, American Nuclear Society, Hinsdale, 1964, pp. 117-136.

Goertz, R. C., et al., An Experimental Head-Controlled TV System to Provide Viewing for a Manipulator Operator, *Proceedings of the 13th Conference on Remote Systems Technology*, American Nuclear Society, Hinsdale, 1965, pp. 53-60.

Goertz, R. C., et al., ANL Mark E4A Electric Master-Slave Manipulator, *Proceedings of the 14th Conference on Remote Systems Technology*, American Nuclear Society, Hinsdale, 1966, pp. 115-123.

Goertz, R. C., J. R. Burnett, and F. Bevilacqua, *Servos for Remote Manipulation*, ANL-5022, 1953.

Goldstein, I. L., and D. A. Schum, Feedback in a Complex Multiman-Machine System, *Journal of Applied Psychology*, Vol. 51, August 1967, pp. 346–351.

Gorka, A. J. Induced Radioactivity and Remote Handling Methods for Accelerators, *Transactions IEEE*, Vol. NS-12, June 1965, p. 656.

Graae, J. E. A., et al., A Radiation Stable Heavy Duty Electromechanical Manipulator, *Proceedings of the Eighth Hot Laboratory and Equipment Conference*, American Nuclear Society, Chicago, 1960, pp. 239-351.

Gregg, L. T., *Man-Machine System Studies*, AD-660647, 1964.

Grenon, M., *Le Travail en Milieu Hostile*, Presses Universitaires de France, Paris, 1968.

Gross, E., Homo Mechanicus, *Science News* Vol. 95, January 11, 1969, pp. 43-45.

Grosson, J. F., Bathyscaphe TRIESTE II Manipulator Arm, *Bureau of Ships Journal*, October 1964 p. 17.

Groth, H. and J. Lyman, Evaluation of Control Problems in Externally Powered Arm Prosthetics, *Orthopedic Prosthetic Appliance Journal*, Vol. 15, June 1961, pp. 174-177.

Groth, H., J. Lyman, and G. Weltman, *Studies in Skilled Myoelectric Control*, AD-633010, 1966.

Gruenberg, E. L., ed., *Handbook of Telemetry and Remote Control*, McGraw-Hill Book Co., New York, 1967.

Haaker, L. W., R. A. Olsen, and D. G. Jelatis, A Gas-Tight Direct-Coupled Mechanical Master-Slave Manipulator for Alpha-Gamma Facilities, *Proceedings of the Tenth Hot Laboratory and Equipment Conference*, American Nuclear Society, Chicago 1962, pp. 153-156.

Hadfield, A., High Speed Forging Control, *Industrial Electronics*, Vol. 3, February 1965, pp. 56-60.

Hancock, R. P., *Mechanical Hand for Forearm Cineplasty*, AD-680953, 1968.

Harrison, L., A Study to Investigate the Feasibility of Utilizing Electrical Potentials on the Surface of the Skin for Control Functions, *Proceedings of the 1964 Seminars on Remotely Operated Special Equipment*, Vol. 1, AEC CONF-640508, 1964, pp. 100-145.

Harrison, S., et al., *Evaluation of Large Scale Visual Displays*, AD-651372, 1967.

Healer, J., et al., *Summary Report on a Review of Biological Mechanisms for Application to Instrumental Design*, Allied Research Association ARA-1025, January 1962, NASA N62-10369.

Healer, J., et al., *A Review of Biological Mechanisms for Application to Instrument Design*, NASA CR-76, 1964.

Heather, A. J., and T. A. Smith, Use of Vital Body Functions to Produce Power for Prosthetic and Orthotic Devices, *Archives of Physical and Medical Rehabilitation*, Vol. 43, June 1962.

Heaton, E. C., J. S. Sweeney, and N. C. Todd, *Studies in Predictor Display Technique*, NASA CR-73068, 1965.

Helberg, L., H. Chiaruttini, and Q. Hartwig, Space Research and the Disabled, *Rehabilitation Record*, Vol. 8, January-February 1967, pp. 2-3.

Heller, R. K., *Accomplishments of the Cable-Controlled Underwater Research Vehicle*, AIAA Paper, 1966.

Henoch, W., and J. Burton, Remote Operations in the SNAP-8 Facility at Atomics International, *Proceedings of the 1964 Seminars on Remote Operated Special Equipment,* Vol. 2, AEO CONF-641120, 1964, pp. 86-97.

Hess, R. A., *The Human Operator as an Element in a Control System with Time-Varying Dynamics,* U.S.A.F. AFFDL-FDCC-TM-65-34, 1965.

Hesson, J. C., M. J. Feldman, and L. Burris, *Description and Proposed Operation of the Fuel Cycle Facility for the Second Experimental Breeder Reactor (EBR II),* AEC ANL-6605, 1963.

Hobbs, L. C., Display Applications and Technology, *Proceedings IEEE,* Vol. 54, December 1966, pp. 1870-1884.

Hoch, R. J., *Control of a Legged Vehicle,* Battelle Northwest Laboratories, Richland, Wash. 1967.

Holeman, J. M., *Optical Instruments for Remotely Controlled Operations,* AEC TID-10074, 1951.

Holeman, J. M., *Design of Periscopes and Remote Viewing Equipment,* General Electric R57GL125, 1957.

Hollister, W. M., An Analytic Measure for the Difficulty of Human Control, *Journal of the Institute of Navigation,* Vol. 20, April 1967, pp. 167-175.

Holmes, A. E., *Space Tool Kit; Survey, Development and Evaluation Program,* Final Report, NASA CR-65267, 1966.

Holzer, F., Programming and Manual Control of Versatile Manipulating Devices, *Proceedings of the 1964 Seminars on Remotely Operated Special Equipment,* Vol. 1, AEC CONF-640508, 1964, pp. 304-324.

Holzer, F. Programming and Manual Control of Versatile Manipulating Devices, *Proceedings of the 1964 Seminars on Remotely Operated Special Equipment,* Vol. 1, AEC CONF-640508, 1964, pp. 304-324.

Homer, G. B., Mobile Manipulator Systems, *Proceedings of the 14th Conference on Remote Systems Technology,* American Nuclear Society, Hinsdale, 1966, pp. 129-137.

Honeywell, Inc., *Feasibility of a Helmet-Mounted Sight as a Control Device,* Rep. S5B-53-1, 1966.

Hormann A., *Designing and Machine Partner—Prospects and Problems,* AD-626173, 1965.

Horn, G. W., Muscle Voltage Moves Artificial Hand, *Electronics,* Vol. 36, October 11, 1963, pp. 34-36.

Howden, G. F., *The Handford Mobile Remote Manipulator System,* AEC HW-76460, 1963.

Howe, R. M., and R. W. Pew, *Development of On-Line Man-Machine System Performance Measurement and Display Techniques,* NASA CR-89964, 1967.

Howell, L. N., and A. M. Tripp, Heavy-Duty Hydraulic Manipulators, *Nucleonics,* Vol. 12, November 1954, pp. 48-49.

Howell, W. C., and G. E. Griggs, *Information Input and Processing Variables in Man-Machine Systems: A Review of the Literature,* AD-230997, 1959.

Howell, W. C., and W. A. Johnston, *Influence of Prolonged Viewing of Large-Scale Displays on Extraction of Information,* AD-660115, 1967.

Huffman, S. A., *Manned Ground Support Equipment*, U.S.A.F. ASD TR 61-430, 1961, pp. 31-42.

Huffman S. A., *Designing for Remote Handling*, Aerojet-General Rep. 2307, 1962.

Hull, H. L., Remote Control Engineering, *Nucleonics*, Vol. 10, November 1952, pp. 34-35.

Hunley, W. H., and W. G. Houck, *Existing Underwater Manipulators*, ASME Paper 65-UNT-8, 1965.

Hunley, W. H., and R. A. Jones, Underwater Remote Handling Systems, *Transactions ANS*, Vol. II, November 1968, p. 617.

Hunt, C. L., and F. C. Linn, *The Beetle, A Mobile Shielded Cab with Manipulators*, SAE Paper 570D; 1962, and *Proceedings of the Tenth Hot Laboratory and Equipment Conference*, American Nuclear Society, Chicago, 1962, pp. 167-184.

Huskey, H. D., Computer Handbook, McGraw-Hill Book Co., New York, 1962.

Huszagh, D. W., Versatile Utility Tool for Hot-Cell Operation, *Proceedings of the Twelfth Conference on Remote Systems Technology*, American Nuclear Society, Hinsdale, 1964, pp. 367-372.

Hyman, A., Utilizing the Visual Environment in Space, *Human Factors*, Vol. 5, 1963, pp. 175-186.

IBM, *Electric Arm Project*, IBM Rep., Endicott, 1950.

Interian, A., and D. A. Kugath, Manipulator Technology—Ready for Space Now, *Astronautics and Aeronautics*, Vol. 7, September 1969, p. 24.

Interian, A., R. H. Blackmer, and W. H. Allen, *Remote Manipulator Spacecraft Systems*, Paper, 20th IAF Congress, Mar del Plata, Argentina, 1969.

Interian, A., et al., *Application of Remote Manipulation to Satellite Maintenance Study*, General Electric DIN 68SD4350, 1968.

Janicke, M. J., and J. C. Carter, A Repairable Nuclear Space Power Plant, *Transactions ANS*, Vol. 7, November 1964 pp. 533-534.

Jeffs, T. W., *New Concepts for the Control of Remote Mechanical Processes*, AEC BNL-302, 1954, pp. 17-24.

Jelatis, D. G., Design Criteria for Heavy-Duty Master-Slave Manipulator, *Proceedings of the Seventh Hot Laboratory and Equipment Conference*, ASME, New York, 1959.

Jelatis, D. G., L. W. Haaker, and R. A. Olsen, A Rugged-Duty Man-Capacity Master-Slave Manipulator, *Proceedings of the Tenth Hot Laboratory and Equipment Conference*, American Nuclear Society, Chicago, 1962, pp. 157-166.

Johnsen, E. G., Teleoperators: Past, Present and Future, *Growth-Change Seminar, Teleoperators and Human Augmentation, Proceedings*, 1969, pp. 38-45.

Johnsen, E. G., Humanoids: The New Remote Systems Technology, *Transactions ANS.*, Vol. 11, November 1968, p. 617.

Johnsen, E. G., *The Case for Localized Control Loops for Remote Manipulators*, IEEE Human Factors Symposium, 1965.

Johnsen, E. G., Telesensors, Teleoperators, and Telecontrols for Remote Operations, *Transactions IEEE*, Vol. NS-13, 1966, pp. 14-21.

Johnsen, E. G. and W. R. Corliss, *Teleoperators and Human Augmentation*, NASA-SP-5047, 1967.

Johnson, A. R., *Processing the Electromonographic Signal, Control of External Power in Upper Extremity Rehabilitation,* NAS/NRC 1352, National Academy of Sciences, Washington, 1966, pp. 208-212.

Johnson, H. C., *Human Factors in the Design of Remote Manipulators,* U.S.A.F. ASD TR 61-430, 1961, pp. 187-192.

Jones, D. G., MRMU in Case of Radioactive Trouble, *Mechanical Engineering,* Vol. 86, May 1963, pp. 29-31.

Jones, D. G., and B. W. Long, *A Test Bed for the Evaluation of Remote Controlled Salvage Concepts,* AIAA Paper 65-525, 1965.

Jones, R. A., Manipulator Systems: A Means For Doing Underwater Work, *Naval Engineering Journal,* February 1968, pp. 107-118.

Judge, J. F., Nuclear Rocketry Hopes Hobbled by Limitations of Remote Handling, *Missiles and Rockets,* Vol. 9, September 18, 1961, pp. 20-22.

Kama, W. N., Human Factors in Remote Handling: A Review of Past and Current Research at the 6570th Aerospace Medical Research Laboratories, *Proceedings of the 1964 Seminars on Remotely Operated Special Equipment,* Vol. 1, AEC CONF-640508, 1964, pp. 198-209.

Kama, W. N., *Effect of Augmented Television Depth Cues on the Terminal Phase of Remote Driving,* U.S.A.F. AMRL TR 65-6, AD-615929, 1965.

Kama, W. N., and R. C. DuMars, *Remote Viewing: A Comparison of Direct Viewing,* 2D and 3D Television, U. S. A. F. AMRL-TDR-64-15, 1964.

Kama, W. N., L. T. Pope, and D. F. Baker, *The Use of Auditory Feedback in Simple Remote Handling Tasks,* U. S. A. F. AMRL-TDR-64-246, 1964.

Kappl, J. J., A Sense of Touch for a Mechanical Hand, S. M. thesis, M. I. T., 1963.

Karinen, R. S. et al., *Summary Report on Mobile Remote Handler,* Sandia Corp. Rep. SCDC-878, 1957.

Karinen, R. S., et al., *Remote Handling Technology and Equipment Investigation,* Phase 1 Report, U. S. A. F. AFSWC-TDR-62-117, AD-290750, 1962.

Karinen, R. S., *Land-Based Remote Handling Background of Underwater Handling Equipment,* ASME Paper 65-UNT-7, 1965.

Karchak, A., and J. R. Allen, *Investigation of Externally Powered Orthotic Devices,* Rancho Los Amigos Hospital, Downey, Calif. 1968.

Keller, G. C., *Man Extension Systems—A Brief Survey of Applicable Techniques,* Goddard Space Flight Center Report X-110-67-618, 1967.

Kelley, C. R., *Manual and Automatic Control,* John Wiley and Sons, New York, 1968.

Kelley, C. R., Predictor Display for Remote Roving Surface Vehicle Control, *Proceedings of the Society for Information Display,* Vol. 10, Winter 1969, pp. 51-56.

Kelley, C. R., *Manual Control Theory and Applications,* AD-449586, 1964.

Kelley, C. R., and P. H. Strudwick, *Manual Control Bibliography,* Dunlap and Associates, Santa Monica, 1964.

Kelley, C. R., Predictor Instruments Look into the Future, *Control Engineering,* Vol. 9, March 1962, pp. 86-90.

Kelley, C. R., et al., *Human Operator Response Speed, Frequency, and Flexibility—A Review, Analysis and Device Demonstration,* NASA CR-874, 1967.

Kelley, C. R., Predictor Displays-Better Control for Complex Manual Systems, *Control Engineering*, Vol. 14, August 1967, pp. 86-90.

Kelley, C. R., *Further Research on the Predictor Instrument*, Tech. Report 252-60-2, Dunlap and Associates, Santa Monica, 1960.

Kelley, C. R., M. B. Mitchell, and P. H. Strudwick, *Applications of Predictor Displays to the Control of Space Vehicles*, Dunlap and Associates, Santa Monica, 1964.

Kelley, C. R., *The Predictor Instrument: 1961*, Dunlap and Associates, Santa Monica, 1962.

Kelley, M. T., and D. J. Fisher, Special Equipment for Analytical Chemistry—Remote Control, *Proceedings of the Fifth Hot Laboratory and Equipment Conference*, 1957.

Klass, P. J., GE Tests Use of Manipulators to Perform Space Repairs, *Aviation Week*, Vol. 89, August 12, 1968, pp. 84-92.

Kleinwachter, H., *The Development of an Anthropomorphous Remotely Operating Machine for the Handling of Radiation Accidents*, IAEA paper M-119/32, Vienna, 1969.

Klepser, W. F., Jr., An Investigation of Some Non-Visual Aids to Remote Manipulation, B.S. thesis, M. I. T., 1966.

Klopsteg, P. E., P. D. Wilson, et al., *Human Limbs and Their Substitutes*, McGraw-Hill Book Co., New York, 1954.

Knowles, W. B., Human Engineering in Remote Handling, U. S. A. F. MRL TDR-62-58, 1962.

Kobrinski, A. E., et al., Problems of Bioelectric Control, *Proceedings First IFAC International Congress, Moscow 1960*, Butterworths, London, 1961, Vol. 2, pp. 619-623.

Kragle, H. A., *Use of Programmable Robots in Glass Handling Operations*, ASTM Paper MS66-541, 1966.

Kral, A. M., Stability Criteria for Feedback Systems with a Time Lag, *Journal of Social and Industrial Applied Mathematics*, A, Vol. 2, 1965.

Krassner, G. N., and J. V. Michaels, *Introduction to Space Communication Systems*, McGraw-Hill Book Co., New York, 1964.

Kreifeldt, J., *Analysis of Predictor Control*, NASA CR-62006, 1964.

Krevitt, R., Remote Maintenance Techniques Proposed for the 200-Gev Accelerator, *Transaction IEEE*, Vol. NS-14, 1967, pp. 997-1003.

Kuehn, R., Display Requirements Assessments for Command and Control Systems, *Information Display*, Vol. 3, December 1966, pp. 43-46.

LaFollette, J. P., and J. L. Dufour, *USAF Shielded Cab Vehicles, Test and Evaluation*, U. S. A. F. AFSWC-TDR-62-137, 1963.

Landers, R. R., *Man's Place in the Dybosphere*, Prentice-Hall, Inc., Englewood Cliffs, 1966.

Landis, D., et al., *Evaluation of Large Scale Visual Displays*, AD-651372, 1967.

Layman, D. C., and G. Thornton, *Remote Handling of Mobile Nuclear Systems*, AEC TID-21719, 1966.

Le Croissette, D. H., and C. E. Chandler, *The Design and Use of the Surface Sampler on the Lunar Spacecraft Surveyor III*, JPL TR 32-1219, 1967.

Leondes, C. T., *Modern Control Systems Theory*, McGraw-Hill Book Co., New York, 1965.

Leslie, J. M., and D. A. Thompson, *Human Frequency Response as a Function of Visual Feedback Display*, Paper at 10th Annual Meeting, Human Factors Society, 1966.

Leslie, J. M., *Effects of Time Delay in the Visual Feedback Loop of a Man-Machine System*, NASA CR-560, 1966.

Leslie, J. M., L. A. Bennigson, and M. E. Kahn, *Predictor Aided Tracking in a System with Time Delay, Performance Involving Flat Surface, Roll, and Pitch Conditions*, NASA CR-75389, 1966.

Levin, A., *Toward a New Concept in Man/Machine Controls*, Paper at Man-Mobility-Survivability Forum, April 11-12, 1967.

Levison, W. H., *Two Dimensional Manual Control System*, NASA SP-128, 1966, pp. 159-180.

Levison, W. H., and J. I. Elkind, *Two-Dimensional Manual Control Systems with Separate Displays*, NASA SP-144, 1967, pp. 29-42.

Levison, W. H. and J. I. Elkind, *Studies of Multivariable Manual Control Systems*, NASA CR-875, 1967.

Lewis, E. E., and H. Stern, *Design of Hydraulic Control Systems*. McGraw-Hill Book Co., New York, 1962.

Li, Y. T., *Man in an Adaptive and Multiloop Control System*, NASA SP-128, 1966, pp. 99-105.

Licklider, J. C. R., Man-Machine Symbiosis, *Transactions IRE*, Vol. HFE-1, March 1960, pp. 4-10.

Licklider, J. C. R., Man-Computer Communication—Introduction, *Communication Processes*, F. A. Geldard, ed., Pergamon Press, New York, 1965.

Licklider, J. C. R., Man-Computer Symbiosis, *Transactions IRE*, Vol. HFE-1, March 1960, pp. 4-10.

Lindgren, N., Human Factors in Engineering Part II: Advanced Man-Machine Systems and Concepts, *IEEE Spectrum*, Vol. 3, April 1966, pp. 62-72.

Ling-Temco-Vought, *Independent Manned Manipulator*, Summary Technical Report, Rep. 00.859, 1966.

Liston, R. A., Walking Machines, *Journal of Terramechanics*, Vol. 1, 1964, pp. 18-31.

Liston, R. A., and R. S. Mosher, *Walking Machine Studies*, Paper 4th Conference U. S./Canadian Section International Terrain Vehicle Systems, no date.

Lockard, R. S., and J. L. Fozard, *The Eye as a Control Mechanism*, U. S. Navy NOTS-TR-1546, 1956.

Lorsch, H. G., Biocontamination Control, *Space/Aeronautics*, Vol. 46, November 1966, pp. 82-91.

Loudon, W. L., Servo Restraint System for Anti-G Protection, *Proceedings of the 1964 Seminars on Remotely Controlled Special Equipment*, Vol. 1, AEC CONF-640508, 1964 pp. 210-222.

Lucaccini, L. F., A. Freedy, and J. Lyman, *Externally Powered Upper Extremity Prosthetic Systems—Studies of Sensory Motor Control*, Report 67-12, University of California, Los Angeles, 1967.

Luxenberg H. R., and R. L. Kuehn, eds., *Display Systems Engineering*, McGraw-Hill Book Co., New York, 1968.

Lyman, J., et al., *Studies in Skilled Myoelectric Control*, AD-633010, 1966.

Lyman, J., and A. Freedy, *"Inhibitory" Control: Concept for a First Model*, In NASA SP-144, 1967, pp. 311-314.

Lyman, J., H. Groth, and G. Weltman, Myoelectric and Mechanical Outputs of Isolated Muscles for Skilled Control Applications. Ergonomics, *Proceedings 2nd IEA Congress*, 1964, pp. 455-462.

Lyman, J., H. Groth, and G. Weltman, Practical Transducer Problems in Electromechanical Control of Arm Prostheses, *Proceedings of the International Symposium on the Application of Automatic Controls in Prosthetic Design*, Opatija, Yugoslavia, 1962.

Lyman, J., and R. L. Smith, The Tracking Loop—A Critical Review of Tracking and Related Sensorimotor Studies, AD-657138, 1966.

Machol, R. E., ed., *System Engineering Handbook*, McGraw-Hill Book Co., New York, 1965.

Magdaleno, R. E., and D. T. McRuer, Effects of Manipulator Restraints on Human Operator Performance, AD-645289, 1966.

Mann, R. W., Recent Progress in the Development of an Electro-Myographically Controlled Limb, *Tech. Digest of 8th Annual IEEE Symposium*, May 1967.

Marquardt, E., Biochemical Control of Pneumatic Prostheses with Special Consideration of Sequential Control, *The Control of External Power in Upper Extremity Rehabilitation*, NAS/NRC Pub. 1352, National Academy of Sciences, Washington, 1966, pp. 20-31.

Martin, O. E., Jr., and R. O. Lowry, *Space Vehicles and Remote-Handling Equipment*, U. S. A. F. ASD TR 61-430, 1961, pp. 79-102.

Martindale, R. L., and W. F. Lowe, *Use of Television for Remote Control: A Preliminary Study*, U. S. A. F. AFSWC-TN-58-12, 1958.

Massachusetts Institute of Technology, Project MAC, Progress Reports, 1965 to date.

Mauro, J. A., *Three-Dimensional Color Television System for Remote Handling Operation*, U. S. A. F. ASD TR 61-430, 1961, pp. 103-168.

Mavor, J. W., Jr., *Alvin, 6000-ft. Submergence Research Vehicle*, Paper, Society Naval Architects and Marine Engineers, 1966.

Mayo, A. M., *Space Exploration by Remote Control*, Paper, International Astronautical Federation, 1964.

Mayo, A. M., *Manned Control—Direct and Remote*, SAE Paper 650811, 1965.

McCandlish, S. G., *A Computer Simulation Experiment of Supervisory Control of Remote Manipulation*, M.I.T. Rep. 9960-2, June 1966.

McCandlish, S. G., R. C. Rosenberg, and T. B. Sheridan, *Investigations in Computer-Aided Instruction and Computer-Aided Controls*, AD-655374, 1967.

McCown, J. J., W. R. Sovereign, and R. P. Larsen, The Use of Commercial Equipment for Analytical Chemistry by Remote Control, *Proceedings of the Seventh Hot Laboratory and Equipment Conference*, ASME, New York, 1959, pp. 219-226.

McCoy, W. K., Jr., and G. G. Frost, *Investigation of "Predictor" Displays for Orbital Rendezvous*, U. S. A. F. AMRL-TR-65-138, 1965.

McCoy, W. K., Jr., and G. G. Frost, *Predictor Display Techniques for On-Board Trajectory Optimization of Rendezvous Maneuvers*, U.S.A.F. AMRL-TR-66-60, 1966.

McCubbin, J. G., and A. S. Bain, Micromanipulator for Use in a Remote Handling Cell, *Proceedings of the Sixth Hot Laboratory and Equipment Conference*, AEC TID-7556, 1958, pp. 237-240.

McLane, J. C., et al., Lunar Receiving Laboratory, *Science*, Vol. 155, February 3, 1967, pp. 525-529.

McLane, R. C., and J. D. Wolf, Symbolic and Pictorial Displays for Submarine Control, NASA SP-128, 1966, pp. 213-228.

McLaurin, C. A., Control of Externally Powered Prosthetic and Orthotic Devices by Musculoskeletal Movement, *The Control of External Power in Upper Extremity Rehabilitation*, National Academy of Sciences, NAS/NRC Publ. 1352, Washington, 1966.

McRuer, D., et al., *A Systems Analysis Theory for Displays in Manual Control*, AD-675983, 1968.

McRuer, D. T., and E. S. Krendel, The Man-Machine Concept, *Proceedings IRE*, Vol. 50, 1962, pp. 1117-1123.

McRuer, D. T., and E. S. Krendel, Dynamic Response of Human Operators, AD-110693, 1957.

McRuer, D. T., and E. S. Krendel, The Human Operator as a Servo System Element, *Journal of the Franklin Institute*, Vol. 267, June 1959, pp. 1-49.

McRuer, D. T., and H. R. Jex, *Systems Analysis Theory of Manual Control Displays*, In NASA SP-144, 1967, pp. 9-28.

McRuer, D. T. and R. E. Magdaleno, *Human Pilot Dynamics with Various Manipulators*, AD-645289, 1966.

McRuer, D. T. et al., *Human Pilot Dynamics in Compensatory Systems—Theory, Models, and Experiments with Controlled Element and Forcing Function Variations*, U. S. A. F. AFFDL-TR-65-15, 1965.

Melton, D. F., Rate Controlled Manipulators, *Proceedings of the 1964 Seminars on Remotely Operated Special Equipment*, Vol. 1, AEC CONF-640508, 1964, pp. 75-93.

Merchant, J., *A New Approach to Space Exploration*, NASA CR-76, 1963.

Merchant J., Oculometer for "Hands-Off" Pointing and Tracking, *Space/Aeronautics*, February 1968, p. 92.

Merchant, J., *The Oculometer*, NASA CR-805, 1967.

Mergler, H. W., and P. W. Hammond, *A Path Optimization Scheme for a Numerically Controlled Remote Manipulator*, NASA CR-62878, 1965.

Miles, L. E., T. C. Parsons, and P. W. Howe, *Force Multiplier for Use with Master Slaves*, University of California UCRL-9662.

Miller, B. P., et al., *Roving Vehicle Motion Control*, NASA CR-92643, 1967.

Miller, G. A., Man-Computer-Interaction, *Communication Processes*, F. A. Geldard, ed., Pergamon Press, New York, 1965.

Miller, J. S., *The Myotron—A Servo-Controlled Exoskeleton for the Measurement of Muscular Kinetics*, Cornell Aeronautical Laboratory Rpt. VO 2401-E-1, 1968.

Milsum, J. H., *Biological Control Systems Analysis*, McGraw-Hill Book Co., New York, 1966.

Minsky, M. L., and S. A. Papert, Research on Intelligent Automata, *Status Report II*, M.I.T. Project MAC, Cambridge, 1967.

Mitchell, M. B., Systems Analysis—The Human Element, *Electro-Technology*, April 1966, pp. 59-72.

Mitchell, M. B., *A Survey of Human Operator Models for Manual Control*, AD-449587, 1964.

Mizen, N. J., Design and Test of a Full-Scale Wearable Exoskeletal Structure, *Proceedings of the 1964 Seminars on Remotely Operated Special Equipment*, Vol. 1, AEC CONF-640508, 1964, pp. 158-186.

Mizen, N. J., The Man Amplifier Concept, *Astronautics & Aeronautics*, Vol. 3, March 1965, pp. 68-71.

Mizen, N. J., *Preliminary Design for the Shoulders and Arms of a Powered, Exoskeletal Structure*, Cornell Aeronautical Laboratory Rep. VO-1692-V-4, 1965.

Mohr, W. C., and C. H. Youngquist, Hinged Arm Polar Manipulator Positioner Mounted on a Radio Controlled Mobile Base, *Proceedings of the Eighth Hot Laboratory and Equipment Conference*, AEC TID-7599, 1960, pp. 230-238.

Moll, J., Some Ideas and Proposals Regarding Standardization Equipment for Hot Laboratories and Remote Control, *Proceedings of the Sixth Hot Laboratory and Equipment Conference*, AEC TID-7556, 1958, pp. 170-182.

Morand, R. F., *Remote Handling*, General Electric APEX-911, 1961.

Morawski, J., *The Role of the Human Factor in Control Systems*, AD-421191, 1963.

Mosher, R. S., *Mechanism Cybernetics*, General Electric Co., Schenectady, 1967.

Mosher, R. S., *Description and Evaluation of "Handyman," Servo Manipulator*, GEL-II, General Electric 59GL235, 1959.

Mosher, R. S., and W. Murphy, *Human Control Factors in Walking Machines*, ASME Human Factors Paper, November 1965.

Mosher, R. S., *Handyman to Hardiman*, SAE Paper 670088, 1967.

Mosher, R. S., An Electrohydraulic Bilateral Servomanipulator, *Proceedings of the Eighth Hot Laboratories and Equipment Conference*, AEC TID-7599, 1960, pp. 252-262.

Mosher, R. S., and B. Wendel, Force Reflecting Electrohydraulic Servomanipulator, *Electro-Technology*, Vol. 66, December 1960, pp. 138-141.

Mosher, R. S., and W. B. Knowles, *Operator-Machine Relationships in the Manipulator*, U. S. A. F. ASD-TR-61-430, 1961, pp. 173-186.

Mosher, R. S., Industrial Manipulators, *Scientific American*, Vol. 211, October 1964, pp. 88-96.

Mosher, R. S., *Dexterity and Agility Improvement*, Paper ASME Underwater Technology Meeting, New London, 1965.

Mosher, R. S., *Design and Fabrication of a Full-Scale, Limited-Motion Pedipulator*, AD-619296, 1965.

Mosher, R. S., J. S. Fleszar, and P. F. Croshaw, *Test and Evaluation of the Limited Motion Pedipulator*, AD-637681, 1966.

Motis, G. M., *Final Report on Artificial Arm and Leg Research and Development*, Northrop Aircraft, Hawthorne, 1951.

Muckler, F. A., and R. W. Obermayer, *Control System Lags and Man-Machine System Performance*, NASA CR-83, 1964.

Murphy, E, Manipulators and Upper-Extremity Prosthetics, *Proceedings of the 1964 Seminars on Remotely Operated Special Equipment*, Vol. 1, AEC CONF-640508, 1694, pp. 380-390.

NASA, *Recent Advances in Display Media*, NASA SP-159, 1968.

NASA, *Second Annual NASA-University Conference on Manual Control*, NASA SP-128, 1966.

NASA, *Third Annual NASA-University Conference on Manual Control*, NASA SP-144, 1967.

NASA, *Vehicle Walks on Varied Terrain Can Assist Handicapped Persons*, NASA Tech. Brief 64-10274, 1964.

NASA, *Advancements in Teleoperator Systems*, NASA SP-5081, 1970.

NASA, *Fifth Annual NASA-University Conference on Manual Control*, NASA SP-215, 1970.

National Research Council, *The Application of External Power in Prosthetics and Orthotics*, Rep. 874, 1961.

National Research Council, *The Control of External Power in Upper-Extremity Rehabilitation*, Rep. 1352, 1966.

Neder, M. J., and C. D. Montgomery, Evolution of Remote Handling Capabilities at NRDS, *Proceedings of the 1964 Seminars on Remotely Operated Special Equipment*, Vol. 1, AEC CONF-640508, 1964, pp. 334-338.

Newman, R. A., Time Lag Consideration in Operator Control of Lunar Vehicles from Earth, *Technology of Lunar Exploration*, C. I. Cummings and H. R. Lawrence, eds., Academic Press, New York, 1963.

Nickel, V. L., Investigation of Externally Powered Orthotic Devices, Final Project Report, Rancho Los Amigos Hospital, Downey, 1964.

Nilsson, N. J., *A Mobile Automaton: An Application of Artificial Intelligence Techniques*, Paper International Joint Conference on Artificial Intelligence, 1969.

North Amercan Aviation, *Optimum Underwater Manipulator Systems for Manned Submersible*, Final Study Report, Rep. C6-65/32, 1966.

Oak Ridge National Laboratory, *Remote Maintenance Tool Catalog No. 58*, AEC ORNL CF-58-6-83, 1958.

Oak Ridge National Laboratory, *Second Information Meeting: Hot Laboratories and Equipment*, AEC ORNL CF-52-10-230, 1952.

Obermayer, R. W., and F. A. Muckler, *Modern Control System Theory and Human Control Functions*, NASA CR-256, 1965.

Obermayer, R. W., W. F. Swartz, and F. A. Muckler, The Interaction of Information Displays with Control System Dynamics in Continuous Tracking, *Journal of Applied Psychology*, Vol. 45, 1961, pp. 369-375.

Oguztoreli, M. N., *Time-Lag Control Systems*, Academic Press, New York, 1966.

Olewinski, W., et al., *Research Study of the Biomedical Aspects of the Proposed Aerospace Environmental Chamber*, U.S.A.F. AEDC-TDR-63-256, 1963, AD-424461, 1963.

Parker, J. W., Holography and Display, *Information Display*, Vol. 3, June 1966, pp. 24-28.

Perry, J., Possibilities of Control Through Surgical Conversion, *The Control of External Power in Upper Extremity Rehabilitation*, NAS/NRC Pub. 1352, National Academy of Sciences, Washington, 1966, pp. 32-34.

Pesch, A. J., and G. R. Simoneau, *Experimental Analysis of Three Types of Position Control for Remote Manipulators*, Electric Boat Co. Report P-417-67-077, New London, 1967.

Pesch, A. J., Behavioral Cybernetic Theory Applied to Problems of Ship Control and Manipulator Operation in Small Submarines, *Journal of Hydronautics*, Vol. 1, July 1967, pp. 35-40.

Peterson, G. E., and J. E. Shoup, Research on Speech Communication and Automatic Speech Recognition, AD-642763, 1966.

Pieper, D. L., *The Kinematics of Maniplators Under Computer Control*, AD-680036, 1968.

Pigg, L. D., *Human Factors in Remote Handling*, U. S. A. F. ASD-TR-61-430, 1961, pp. 3-8.

Pigg, L. D., *Human Engineering Principles of Design for In-Space Maintenance*, U. S. A. F. ASD-TR-61-629, 1961.

Pingle, K. K., J. A. Singer, and W. M. Wichman, *Computer Control of a Mechanical Arm through Visual Input*, IFIP Paper, 1968 Congress, Edinburgh, 1968.

Pizzicara, D. J., *Computers and Displays/Controls State-of-the-Art Technology Studies*, AD-631663, 1966.

Pollack, I., and L. Ficks, Information of Elementary Multi-Dimensional Auditory Displays, *Journal of the Acoustical Society of America*, Vol. 26, 1954, pp. 155-158.

Poole, H. H., *Fundamentals of Display Systems*, Spartan Books, Washington, 1966.

Porges, I., Famous Robots of the Past, *Science Digest*, Vol. 41, March 1957, pp. 13-16.

Potts, C. W., G. A. Forster, and R. H. Maschhoff, Transistorized Servo System for Master-Slave Electric Manipulators, *Proceedings of the Ninth Hot Laboratory and Equipment Conference*, American Nuclear Society, Chicago, 1961, pp. 154-160.

Price, H. E., and B. J., Tabachnick, *A Descriptive Model for Determining Optimal Human Performance in Systems*, Vol. III, NASA CR-878, 1968.

Rader, P. J., *Variable Flexibility Tether System*, AD-677687, 1968.

Raleigh, H. D., compiler, *Remote Control Equipment (A Literature Search)*, AEC TID-3549, 1960.

Raphael, B., *Programming a Robot*, Stanford Research Institute, Palo Alto, 1967.

Rarich, T. D., *Development of SCM-1 A System for Investigating the Performance of a Man-Computer Supervisory Controlled Manipulator*, M.IT. Report DSR-9991-3, Cambridge, 1966.

Rawson, A. J., *Remote Control of Biologically Hazardous Laboratory Manipulation: A Feasibility Study*, U. S. Army BWL-23, 1960.

Regan, J. J., *Tracking Performance Related to Display Control Configurations*, U.S. Navy NAVTRADEVEN 322-1-2, January 1959.

Reswick, J. B., *Synthetic Muscle Motor Development*, Case Institute of Technology Rep. EDC 4-61-1, 1961.

## Bibliography

Reswick, J. B., and L. Vodovnik, External Power in Prosthetics and Orthotics, an Overview, *Artificial Limbs*, Vol. 11, Autumn 1967, pp. 5-21.

Richards, P., Brookhaven Mechanical Manipulator Model No. 3., *Proceedings of the Fourth Hot Laboratory and Equipment Conference*, 1956, pp. 26-34.

Richards, P., and A. C. Rand, *Some Pieces of Equipment Developed at Brookhaven National Laboratory*, BNL-2183, 1954.

Richter, W., M. Ellefsplass, and R. A. Horne, Telemanipulation in a Big Accelerator, *Industries Atomiques*, Vol. 11, 1967, pp. 77-85.

Ring, F., Remote-Control Handling Devices for Conducting Research and Development Work Behind Shielding Walls of Hot Laboratories, *Mechanical Engineering*, Vol. 78, 1956, pp. 828-831.

Ring, F., compiler, *Sixth Hot Laboratories and Equipment Conference*, AEC TID-7556, 1958.

Robinson, G. H., The Human Controller as an Adaptive, Low Pass Filter, *Human Factors*, Vol. 9, April 1967, pp. 141-147.

Rock Island Arsenal, *The Peripod*, Rock Island, November 1967.

Rock Island Arsenal, *Critique on the Single-Operator-Controlled Quadruped Walking Machine*, Rock Island, 1968.

Rohm & Haas Co., *An Evaluation of Safety Devices for Laboratories Handling Explosive Compounds*, AD-250902, 1961.

Rosen, C., and N. Nilsson, eds., *Application of Intelligent Automata to Reconnaisance*, U. S. A. F. RADC-TR-67-657, 1968.

Rosen, C. A., Machines That Act Intelligently, *Science Journal*, Vol. 4, October 1968, p. 109.

Rothchild, R. A., *Design of an Externally Powered Artificial Elbow for Electromyographic Control*, SM thesis, M.I.T., Cambridge, 1965.

Rouze, E. R., et al., Surveyor Surface Sampler Instrument, JPL TR-32-1223, 1968.

Santschi, W. R., ed., *Manual of Upper Extremity Prosthetics*, 2nd ed., University of California, Los Angeles, 1958.

Savant, C. J., Jr., *Control System Design*, McGraw-Hill Book Co., New York, 1964.

Scott, R. N., Myoelectric Control of Prostheses. *Archives of Physical Medicine and Rehabilitation*, Vol. 47, March 1966, pp. 174-181.

Scott, R. N., Myo-electric Control, *Science Journal*, Vol. 2, March 1966.

Seale, L. M., W. E. Bailey, and W. E. Powe, *Study of Space Maintenance Techniques*, U.S.A.F. ASD TDR-62-931, AD 406776, 1962.

Seale, L. M., and P. N. Van Schaik, *Space Extravehicular Operations, A Review of the Requirements and Alternate System Approaches*, Paper, International Astronautical Congress, Madrid, 1966.

Sebesta, H. R., *Analysis and Design of Optimal Control Systems for Dynamical Processes with Time Delays*, Ph. D. thesis, Texas University, Austin, 1966.

Seeley, H. F., and J. C. Bliss, Compensatory Tracking with Visual and Tactile Displays, *Transactions IEEE*, Vol. HFE-7, June 1966, pp. 87-90.

Seidenstein, S., and A. G. Berbert, Jr., *Manual Control of Remote Manipulators: Experiments Using Analog Simulation*, AD-638500, 1966.

Selwyn, D., *Head-Mounted Inertial Servo Control for Handicapped*, IEEE Paper, 6th Annual Symposium, Prosthetics Group, 1965.

Senders, J. W., The Human Operator as a Monitor and Controller of Multidegree of Freedom Systems, *Transactions IEEE*, Vol. HFE-5, September 1964, pp. 2-5.

Senders, J. W., Man's Capacity to Use Information from Complex Displays, *Information Theory in Psychology*, H. Questler, ed., The Free Press, Glencoe, Illinois, 1955.

Serendipity Associates, *A Descriptive Model for Determining Optimal Human Performance in Systems*, NASA CR-876, Vol. 1, 1968; NASA CR-877, Vol. 2, 1968; NASA CR-878, Vol. 3, 1968; NASA CR-879, Vol. 4, 1967.

Sheridan, T. B., *Time-Variable Dynamics of Human Operator Systems*, AFCRC-TN-60-169, AD-237045, 1960.

Sheridan, T. B., and W. R. Ferrell, Remote Manipulative Control with Transmission Delay, *Transactions IEEE*, Vol. HFE-4, September 1963, pp. 25-28.

Sheridan, T. B., and W. R. Ferrell, *Measurement and Display of Control Information*, NASA CR-93853 and other Progress Reports, 1967.

Sheridan, T. B., On Precognition and Planning Ahead in Manual Control, *Proceedings of the Fourth National Symposium on Human Factors in Electronics*, 1963.

Sheridan, T. B., et al., Control Models of Creatures Which Look Ahead, *Proceedings of the Fifth National Conference on Human Factors in Electronics*, 1964.

Sheridan, T. B., The Human Operator in Control Instrumentation, *Progress in Control Engineering*, Vol. 1, 1962, pp. 143-187.

Sheridan, T. B., *Studies of Adaptive Characteristics of the Human Controller*, AD-297367, 1962.

Sheridan, T. B., Three Models of Preview Control, *Transactions IEEE*, Vol. HFE-7, June 1966, pp. 91-102.

Sheridan, T. B., and W. R. Ferrell, *Supervisory Control of Manipulation*, In NASA SP-144, 1967, pp. 315-323.

Sheridan, T. B., and W. R. Ferrell, *Functional Extension of the Human Hands*, Progress Reports, NASA CR-69856, NASA CR-70782, 1965.

Shigley, J. E., *The Mechanics of Walking Vehicles*, U. S. Army ATAC RR LL-71, 1960.

Shissler, W. C., Jr., *ANPD Remote Handling Design and Data Book*, General Electric Rep. DC-58-9-112, 1957.

Siegel, A. I., and J. J. Wolf, A Technique for Evaluating Man-Machine System Designs, *Human Factors*, Vol. 2, March 1961, pp. 18-28.

Sinaiko, H., ed., *Selected Papers on Human Factors in the Design and Use of Control Systems*, Dover Publications, New York, 1961.

Skachkov, A. N., Bioelectrically Controlled Artificial Hands in *External Control of Human Extremities*, Yugoslav Committee for Electronics and Automation, Belgrade, 1967, pp. 97-101.

Skolnick, A., Stability and Performance of Manned Control Systems, *Transactions IEEE*, Vol. HFE-7, June 1966, pp. 115-124.

Slowick, J., *Power-Driven Articulated Dummy*, IIT Research Institute, Final Report Project No. K6051, 1965.

Smith, K. U., *Delayed Sensory Feedback and Behavior*, W. B. Saunders Co., Philadelphia, 1962.

Smith, K. U., Cybernetic Foundations of Physical Behavioral Science, *Quest*, Vol. 8, 1963, pp. 26-89.

Smith, R. L., et al., Effects of Display Magnification, Proprioceptive Cues, Control Dynamics and Trajectory Characteristics on Compensatory Tracking Performance, *Human Factors*, Vol. 8, October 1966, pp. 427-434.

Smith, W. M., et al., Delayed Visual Feedback and Behavior, *Science*, Vol. 132, 1960, pp. 1013-1014.

Snelson, R., A. Karchak, and V. L. Nickel, Application of External Power in Upper Extremity Orthotics, *Orthopedic and Prosthetic Appliance Journal*, Vol. 15, December 1961, pp. 345-348.

Spooner, M. G., and C. H. Weaver, An Analysis and Analogue Computer Study of a Force-Reflecting Positional Servomechanism, *Transactions AIEE*, Vol. 74, Part 2, January 1956, pp. 384-387.

Stang, L. G., Jr., *Articulated Tongs*, AEC TID-5280, 1965, pp. 35-45.

Stang, L. G., Jr., compiler, *Hot Laboratory Equipment*, 2nd ed., Government Printing Office, Washington, 1958.

Stang, L. G., Jr., Rectilinear Manipulator BNL Model 4, *Proceedings of the Seventh Hot Laboratory and Equipment Conference*, ASME New York, 1959, pp. 169-176.

Stapleford, R. L., D. T. McRuer, and R. Magdaleno, *Pilot Describing Function Measurements in a Multiloop Task*, NASA CR-542, 1966.

Steele, R. V., and H. B. Thomas, *A Supplementary Remote Control Manipulator*, AEC LRL-129, 1954.

Stevenson, C. E., et al, Maintenance and Repair of Contaminated Equipment for the EBR-II Fuel Cycle Facility, *Proceedings of the 14th Conference on Remote Systems Technology*, American Nuclear Society, Hinsdale, 1966, pp. 149-155.

Stevenson, D. A., *Engineering an Artificial Arm*, Paper at 1967 Congress of Canadian Engineers, 1967.

Streechon, G. P., A Remotely Maintainable Rectilinear Manipulator, *Proceedings of the Eighth Hot Laboratory and Equipment Conference*, AEC TID-7599, 1960, pp. 277-296.

Strickler, T. G., Design of an Optical Touch Sensor for a Remote Manipulator, S.M. thesis, M.I.T., 1966.

Sullivan, G., et al., *Myoelectric Servo Control*, U.S.A.F. ASD-TDR-63-70, 1963.

Sullivan, G., et al., *Myoelectric Servo Control*, AD-410898, 1963.

Summers, L. G., and K. Ziedman, *A Study of Manual Control Methodology with Annotated Bibliography*, NASA CR-125, 1964.

Sutherland, I. E., Computer Outputs and Inputs, *Scientific American*, Vol. 215, September 1966, pp. 86-96.

Sutherland, N. S., Machines Like Men, *Scientific Journal*, Vol. 4, October 1968, pp. 44-49.

Sweeney, J. S., N. C. Todd, and E. C. Heaton, Studies in Predictor Display Technique, NASA CR-73068, 1965.

Szego, G. C., and J. E. Taylor, eds., *Space Power Systems Engineering*, Academic Press, New York, 1966.

Talbot, J. E., Programmable Industrial Robots, *ISA Journal*, Vol. 13, September 1966, pp. 31-36.

Taylor, F. V., and W. D. Garvey, The Limitations of a "Procrustean" Approach to the Optimization of Man-Machine Systems, *Ergonomics*, Vol. 2, February 1959, pp. 187-194.

Taylor, F. V., and H. P. Birmingham, *A Human Engineering Approach to the Design of Man-Operated Continuous Control Systems*, U. S. Navy NRL-4333, 1954.

Taylor, R. J., *A Digital Interface for the Computer Control of a Remote Manipulator*, NASA CR-80843, 1966.

Thomas, P. G., Large-Screen Displays, *Space/Aeronautics*, Vol. 47, May 1967, p. 82.

Thomas, R. E., and J. T. Tou, Evolution of Heuristics by Human Operators in Control Systems, *IEEE International Convention Record*, Vol. 15, Pt. 9, pp. 179-192.

Thomas, R. E., and J. T. Tou, *Human Decision-Making in Manual Control Systems*, NASA SP-128, 1966, pp. 325-334.

Thompson, W. M., Force-Reflecting Servomechanism with Signal Relay, *Transactions ANS*, Vol. 11, June 1968, p. 361.

Thompson, W. M., A. G. Vacroux, and C. H. Hoffman, *Application of Pontryagin's Time Lag Stability Criterion to Force-Reflecting Servomechanisms*, Paper at Ninth Joint Automatic Control Conference, Ann Arbor, 1968.

Thompson, W. M., and R. C. Goertz, Master-Slave Servo-Manipulator, Model 2, *Proceedings of the Fourth Hot Laboratory and Equipment Conference*, 1956, pp. 1-10.

Tilton, H. B., Principles of 3-D CRT Displays, *Control Engineering*, Vol. 13, February 1966, pp. 74-78.

Tobey, W. H., R. T. French, and D. M. Adams, *Experimental Material Handling Device, Summary Report*, Martin Marietta Rpt. MCR-69-414, 1969.

Tolliver, R. L., USAF Exploratory Development of Remote Handling Equipment, *Proceedings of the 1964 Seminars on Remotely Operated Special Equipment*, Vol. 1, AEC CONF-640508, 1964, pp. 327-331.

Tomavic, G., and G. Broni, An Adaptive Artificial Hand, *Transactions IRE*, Vol. AC-7, April 1962, pp. 3-9.

Tomovic, R., Outline of a Control Theory of Prosthetics, *Automatic and Remote Control*, V. Broida, ed., Butterworths, London, 1964, pp. 449-453.

Tomovic, R., and R. B. McGhee, A Finite State Approach to the Synthesis of Bioengineering Control Systems, *Transactions IEEE*, Vol. HFE-7, June 1966, pp. 65-69.

Tomovic, R., Control Theory and Signal Processing in Prosthetic Systems, *The Control of External Power in Upper Extremity Rehabilitation*, NAS/NRC 1352, National Academy of Sciences, Washington, 1966, pp. 221-226.

Tou, J. T., *Digital and Sampled-data Control Systems*, McGraw-Hill Book Co., New York, 1959.

Truxal, J. G., *Automatic Feedback Control System Synthesis*, McGraw-Hill Book Co., New York, 1955.

Truxal, J. G., *Control Engineers' Handbook*, McGraw-Hill Book Co., New York, 1958.

Tustin, A., The Nature of the Operator's Response in Manual Control and Its Implications for Controller Design, *Journal of the Institute of Electrical Engineers*, Vol. 94, 1947.

Tustin, A., ed., *Automatic and Manual Control*, Academic Press, New York, 1964.

U.S. Government, *The Control of External Power in Upper Extremity Rehabilitation*, National Research Council Rep. 1352, 1966.

U.S. Government, *The Application of External Power in Prosthetics and Orthotics*, National Research Council Rep. 874, 1961.

Vallerie, L. L., Displays for Seeing Without Looking, *Human Factors*, Vol. 8, December 1966, pp. 507-513.

Van Dam, A., Computer Driven Displays and Their Use in Man/Machine Interaction, *Advances in Computers*, Vol. 7, 1966, pp. 239-290.

Verplank, W. L., Symbolic and Analogic Command Hardware for Computer-Aided Manipulation, M.S. Thesis, M.I.T., Cambridge, 1967.

Vertut, J., New Types of Heavy Manipulators, *Proceedings of the Tenth Hot Laboratory and Equipment Conference*, American Nuclear Society, Chicago, 1962, pp. 185-194.

Vinograd, S. P., *Medical Aspects of an Orbiting Research Laboratory*, NASA SP-86, Washington, 1966.

Vivian, C. E., W. H. Wilkins, and L. L. Haas, Remotely Operated Service Module for Maintenance of Orbital Systems, *Proceedings of the 12th Conference on Remote Systems Technology*, American Nuclear Society, Hinsdale, 1964, pp. 89-104.

Vivian, C. E., W. H. Wilkins, and L. L. Haas, Advanced Design Concepts for a Remotely Operated Manipulator System for Space Support Operations, *Proceedings of the 1964 Seminars on Remotely Operated Special Equipment*, Vol. 1, AEC CONF-640508, 1964, pp. 248-299.

Vodovnik, L., et al., *Some Topics on Man-Machine Communication in Orthotic and Prosthetic Systems*, Case Institute of Technology, EDC Rpt. 4-67-16, 1967.

Wagman, I. H., and D. S. Pierce, Electromyographic Signals as a Source of Control, *The Control of External Power in Upper Extremity Rehabilitation*, National Academy of Sciences, NAS/NRC Publ. 1352, Washington, 1966, pp. 35-56.

Walker, J. A., Stability of Feedback Systems Involving Time Delays and a Time-Varying Non-Linearity, *Journal of Control*, Vol. 6, October 1967, pp. 365-372.

Wargo, M. J., Delayed Sensory Feedback in Visual and Auditory Tracking, *Perceptual and Motor Skills*, Vol. 24, February 1967, pp. 55-62.

Wargo, M. J., et al., *Human Operator Response Speed, Frequency, and Flexibility*, NASA CR-874, 1967.

Wargo, M. J., et al., Muscle Action Potential and Hand Switch Disjunctive Reaction Times to Visual, Auditory, and Combined Visual-Auditory Displays, *Transactions IEEE*, Vol. HFE-8, September 1967, pp. 223-226.

Waring, W., and V. L. Nickel, *Investigation of Myoelectric Control of Functional Braces*, Rancho Los Amigos Hospital, Downey, Calif., 1968.

Warner, J. D., *A Fundamental Study of Predictive Display Systems*, NASA CR-1274, 1969.

Wasserman, W. L., Human Amplifiers, *International Science and Technology*, October 1964, pp. 40-48.

Webb, P., ed., *Bioastronautics Data Book*, NASA SP-3006, Washington, 1964.

Weissenberger, S., and T. B. Sheridan, Dynamics of Human Operator Control Systems Using Tactile Feedback, *Journal of Basic Engineering*, Vol. 84, 1962, pp. 297-301.

White, L. E., Remote Handling Requirements for NERVA (E-MAD & ETS-1), *Proceedings of the 1964 Seminars on Remotely Operated Special Equipment*, Vol. 2, AEC CONF-641120, 1964, pp. 9-35.

Whitney, D. E., *State Space Models of Remote Manipulation Tasks*, Ph. D. Thesis, M.I.T., Cambridge, 1968.

Wierwille, W. W., and G. A. Gayne, *Time Varying and Nonlinear Models of Human Operator Dynamics*, NASA SP-128, 1966, pp. 107-108.

Wiesener, R. W., The Minotaur I Remote Maintenance Machine, *Proceedings of the Eleventh Hot Laboratory and Equipment Conference*, American Nuclear Society, Hinsdale, 1963, pp. 197-210.

Williams, W. L., A. G. Berbert, Jr., and F. A. Maher, *Control of Remote Manipulator Motion*, Ritchie, Inc., Dayton, 1966.

Williams, W. L., and A. G. Berbert, Jr., *Combined Control of Remote Manipulator Translation and Rotation Motion*, Ritchie, Inc., Dayton, 1967.

Wilson, A. B. K., Control Methods with Mechanical and Electromyographic Inputs, *The Control of External Power in Upper Extremity Rehabilitation*, NAS/NRC Publ. 1352, National Academy of Sciences, Washington, 1966, pp. 57-60.

Wilson, K., Discussion of Pneumatically Actuated Master Slave Manipulator, *Proceedings of the 1964 Seminars on Remotely Operated Special Equipment*, Vol. 2, AEC CONF-641120, 1964, pp. 63-66.

Wilts, C. H., *Principles of Feedback Control*, Addison-Wesley Publishing Company, Reading, 1960.

Wirta, R. W., *EMG Control of External Power*, ASME Paper 65-WA/HUF-3, 1965.

Woodson, W. E., and D. W. Conover, *Human Engineering Guide for Equipment Designers*, University of California Press, Berkeley, 1964.

Wooldridge, D. E., *Mechanical Man*, McGraw-Hill Book Co., New York, 1968.

Wulfeck, J. W., and L. R. Zeitlin, Human Capabilities and Limitations in *Psychological Principles in System Development*, R. M. Gagne, ed., Holt, Rinehart and Winston, New York, 1962.

Wulff, J. J., et al., *A Descriptive Model for Determining Optimal Human Performance in Systems*, NASA CR-876, 1968.

Yilmaz, H., *A Program of Research Directed Toward the Efficient and Accurate Machine Recognition of Human Speech*, NASA CR-80020, 1966.

Yilmaz, H., *A Program of Research Directed Toward the Efficient and Accurate Machine Recognition of Human Speech, A Theory of Speech Perception*, Final Report, NASA CR-86027, 1967.

Young, L. R., et al., *The Adaptive Dynamic Response Characteristics of the Human Operator in Simple Manual Control*, NASA TN D-2255, 1964.

Young, L. R., and L. Stark, *Biological Control Systems—A Critical Review and Evaluation*, NASA CR-190, 1965.

Zacharias, E. H., Jr., *Development and Evaluation of Keyboard Programming to Input Geometrical Information for Supervisory Manipulation*, thesis, M.I.T., Cambridge, 1967.

Ziebolz, H., and H. M. Paynter, Possibilities for a Two-Time-Scale Computing System for Control and Simulation of Dynamic Systems, *Proceedings of the National Electronics Conference*, Vol. 9, 1953, pp. 215-223.

# INDEX

Actuator subsystem, 4, 5, 37–38, 45, 46, 48, 49, 70, 171–220
Adaptive control, 73–74, 79, 85, 88–90
AEC, 3, 6, 10, 24, 25, 26, 28, 35, 61, 62, 126, 142, 153, 154, 184, 207; *see also* ANL; Brookhaven National Laboratory; Los Alamos; Oak Ridge National Laboratory
Aerospace applications, 11, 13, 14, 15–19, 47, 64
Alderson, S., 209, 210
Allen, J., 8
Alvin, I., 20, 150, 151, 192
American Car & Foundry, 198
American Machine & Foundry, 185, 186, 187, 198
Analog control, 96–97, 111, 119–120
Anderson, V. C., 20, 199
ANL, 3, 6, 7, 10, 17, 18, 29, 35, 38, 49, 60, 66, 67, 70, 79, 91, 98, 108, 120, 123, 125, 126, 150, 156, 157, 158, 172, 183, 184, 185, 188, 200, 206, 211, 212, 213, 214
Anthropomorphism, 37, 38, 45, 76, 79, 80, 83, 108–109, 111, 112, 117, 118, 121, 126, 129, 130, 147, 162, 164, 165, 174; *see also* Man-machine interface
Argonne National Laboratory, *see* ANL
Artificial intelligence, 74, 85, 88–90
Artificial limbs, *see* Prosthetics
Ashera, 20
Atomic Energy Commission, *see* AEC
Attitude control subsystem, 5, 40, 53, 69–71
Autec I, 20, 61

Autonetics, 198

Bat, 28
Battelle Northwest Laboratories, 89, 128
Beaver, 20, 150
Beetle, 20
Berbert, A. G., Jr., 193
B. F. Goodrich, 214
Bilateral teleoperator, 38, 98–99, 111, 124, 127, 172, 173, 183, 204; *see also* Force feedback
Bliss, J. C., 7, 160, 169, 170
Bradley, W. E., 3, 7, 13, 129, 205
Brookhaven National Laboratory, 28, 29, 126, 176, 183, 184, 212, 214
Burnett, J. R., 98

CAM, 3, 29
Case Western Reserve, 7, 85, 88, 108, 142, 144
Central Research Laboratories, *see* CRL
Chatten, J., 8
Clark, J. W., 3
Communications subsystem, 4, 5, 39, 42, 48, 51, 54–57, 58
Compliance, 46, 82, 126, 180, 208
Computer Image Corp., 164
Computers, 7, 74, 167
Computer subsystem, 5, 39, 51, 220; *see also* Preview control; Supervisory control
Control Data Corp., 157
Control subsystem, 4, 39, 50, 73–144, 220
Control theory, 83–99
Cornell Aeronautical Laboratory, 7, 29, 30, 68, 130, 202

249

250  Index

Cousteau, J., 20
Crawford, B., 118
CRL, 122, 185, 188, 194
Cudworth, A. L., 211
CURV, 23

Deepstar, 20
Degrees of freedom, 46, 79, 80, 81, 116, 121, 124, 136, 146, 156, 160, 165, 174, 180, 183, 193, 201–202, 208, 212
Department of Defense, 30, 58, 97, 128, 142; see also U.S. Army, etc.
Design philosophy, 42, 44–48
Discoverer I, 192
Disney, W., 35, 38
Displays, 45, 47, 59, 74–75, 79, 162–170
Diving Saucer SP–300, 20, 21, 191, 192, 216
DOWB, 20
DSRV–1, 23, 28, 61, 67, 119
Dunlap and Associates, 168

EBR II, 25, 157
Edgerton, Germeshausen, and Grier, 149
Electric Boat Division, 193, 196, 199
Electromyography, see EMG
EMG, 32, 50, 77, 78, 110, 111, 138–141, 206, 211
Environment control subsystem, 40, 53, 71, 76, 220
Exoskeletons, see Hardiman; Man amplifier; Walking machines

Feedback, 86–88; see also Force feedback
Ferrell, W. R., 74, 91
Flatau, 126, 212
Force feedback, 45, 76, 98, 121, 127, 162, 166, 168–170, 207, 208, 211; see also Bilateral teleoperator; Master-slave
Forging manipulator, 33, 62, 65, 190

General Electric, 6, 7, 15, 17, 19, 24, 30, 57, 61, 63, 80, 97, 99, 126, 128, 129, 130, 131, 132, 155, 184, 192, 195, 200, 201, 202, 203, 204, 205, 209, 212
General Mills, 6, 7, 118, 192, 208
General Motors, 168
Giannini Controls, 215
Glimcher, M. J., 211
Goertz, R. C., 3, 6, 7, 9, 98, 120, 156, 183, 211

Handyman, 58, 80, 114, 126, 127, 132, 160, 190, 200–201
Hardiman, 30, 68, 99, 129, 130, 131, 133, 190, 192, 201, 202–205
Harvard University, 211
Hoch, R. J., 89, 128
Honeywell, 132
Houck, W. G., 22
Hughes Tool Co., 22
Human factors, 75–78; see also Anthropomorph Interfaces; Man-machine interface
Human transfer function, 75, 87
Hunley, W. H., 22, 192
Hydroman, 190, 192

IBM, 209, 210
Indexing, 81, 126, 179, 212–213
Institute for Defense Analyses, 205
Institutes for the Achievement of Human Potential, 32
Interfaces, 5, 37–43; see also Man-machine interface

Jelatis, D., 178
Jex, H. R., 163
Johnsen, E. G., 29
Joystock, 111, 116–119, 207, 208
JPL, 112

Kama, W. N., 118, 158
Karchak, A., Jr., 8
Kelley, C. R., 116, 163, 167
Klepser, W. F., 158
Koelsch Corp, 197

Liberty Mutual Insurance Co., 211
Ling-Temco-Vought, 17, 18, 60, 66, 67, 70, 125
Liston, R. A., 29, 206
Litton Industries, 195
Lockheed Missiles & Space Co., 23
Los Alamos, 208, 209
Lyman, J., 94, 96, 141

MAIS project, 30, 202
Man amplifier, 2, 14, 29–30, 49, 64, 68, 76, 12 130–132, 202–205; see also Hardiman
Man-machine interface, 11, 75–80, 100–110, 2
Mann, R. W., 211
Marshall Space Flight Center, 125, 208, 212
Masher, 28

Master-slave, 3, 6, 7, 9, 10, 20, 24, 25, 28, 32, 37, 42, 44, 49, 50, 51, 53, 54, 55, 58, 64, 65, 67, 69, 72
  control, 119, 120–128
  definition, 38
  electric, 6, 123–125, 158, 172, 181, 206, 207, 211–214
  electrohydraulic, 200–201; *see also* Handyman
  hydraulic, 191–192
  mechanical, 121–123, 176, 181, 184–188
Melton, D. F., 207
Minotaur, 153, 158, 208, 209
Minsky, M., 199
M.I.T., 85, 89, 94, 96, 108, 134, 135, 142, 144, 160, 161, 199, 211
Mobot, 22, 61, 153
Morrison, R. A., 7
Mosher, R. S., 3, 128, 206
Moulton, S., 7
MRMU, 28, 58, 61, 62, 69, 72
Myoelectricity, *see* EMG

NASA, 7, 17, 18, 21, 57, 60, 62, 70, 86, 89, 90, 91, 100, 112, 114, 132, 142, 160, 166, 168, 207
National Aeronautics and Space Administration, *see* NASA
Norden Division, United Aircraft Corp., 15
North American Aviation, 20, 150, 154, 199
Northern Electric Co., 137
NRDS, 26, 62, 63, 154, 178
Nuclear applications, 14, 23–29, 47, 64
Nuclear Rocket Development Station, *see* NRDS

Oak Ridge National Laboratory, 149, 192
Oceanographic Engineering Corp., 154
Orthotics, 31, 77, 135–141

PaR, 26, 114, 194, 208
PaR–1, 26, 27, 28, 35, 61
Payne, J., 6
Paynter, H. M., 167
Pesch, A., 118
Philco Corp., 156, 158
Position control, *see* Analog control
Power subsystem, 5, 39–40, 52, 62–69, 220
Predictor displays, 79, 163, 166–167
Preprogrammed machines, 73, 79, 83, 84–86, 89, 115, 120, 206

Preview control, 57, 59, 79, 90
Programmed and Remote Systems Corp., *see* PaR
Propulsion subsystem, 5, 39, 52, 59–62
Prosthetics, 14, 31–32, 47, 51, 58, 62, 64, 67, 68–69

Quickening, 79, 167

Rancho Los Amigos Hospital, 8, 137
Recoverer I, 22, 191
Rectilinear teleoperator, 38, 79; *see also* Unilateral teleoperator
Replica control, *see* Analog control
Ritchie, Inc., 97
RUM, 21–22, 61, 73, 74, 79, 108, 142–144, 153, 192

Scripps Institution of Oceanography, 18, 20, 61, 197, 199, 200
Sea Cliff, 20
Seidenstein, S., 193
Sensor subsystem, 4, 5, 38–50, 145–170, 220
Serpentuator, 37, 173, 208–209
Shell Oil Co., 22
Sheridan, T. B., 74, 89, 94, 128, 134, 142, 160
Shigley, J. E., 7, 206
Smith, K. U., 93
SNAP, 26, 154
Sonar, 147, 158, 159
Space-General Corp., 7, 128, 206
Spatial Correspondence, 42, 44–45, 79, 80, 121, 126, 130, 183, 193
Stanford Research Institute, 135, 142–144, 160, 169
Stanford University, 91, 168
Structure subsystem, 40, 53, 72, 220
Supervisory control, 51, 57, 74, 79, 83, 84–86, 90, 106, 107, 111, 120, 128, 141–144, 147, 210
Surveyor surface sampler, 11, 14, 84, 85, 112–113, 116, 168
Switch controls, 111–116, 133–138

Tactual feedback, 168–170
Telechirics, 3
Telefactor, 3
Teleoperators, applications, 9–10, 11–36
  definition, 1, 2
Telepuppet, 3
Television, 57, 152, 153–158, 160

Terminal devices, 188, 215–218
Thompson, W. H., 91
Thresher, 20, 23
Time-delay problem, 13, 18–23, 39, 47, 79, 84, 90–92, 114, 115, 128, 163
Tongs, 24, 182–183
Tools, *see* Terminal devices
Touch sense, 159–162; *see also* Force feedback
Transportation applications, 14, 29–31
Trieste, 7, 20, 192
Turtle, 20

Undersea applications, 14, 47, 64, 190–200
Unilateral teleoperators, 38, 39, 44, 49, 50, 55, 58, 172, 173, 176, 177, 183–184, 192–200
  control, 96–97, 111, 117, 118
  definition, 38, 63
United States Air Force, 7, 15, 28, 81, 97, 151, 152, 160, 168, 200, 202
United States Army, 7, 29, 32–33, 68, 129, 189, 202
United States Navy, 7, 23, 30, 68, 167, 192, 202
United States Veterans Administration, 209
University of California at Los Angeles, 94
UNUMO, 22

Visual correspondence, 79

Walking machines, 3, 22, 29–31, 49, 52, 53, 64, 69–70, 72, 76, 89, 128–132, 205–206, 219; *see also* Hardiman; Man amplifier
Walter Reed Army Medical Center, 138
Wargo, M. J., 91
Watson, T. J., 209
Westinghouse Electric Co., 119, 195, 196
Whipple, F. L., 3
Whitney, D. E., 94
Wiener, N., 6
Windows, hot-cell, 149–152

Ziebolz controller, 167